HOW AND WHERE TO FIND GOLD

VERNE H. BALLANTYNE

ARCO PUBLISHING, INC.
NEW YORK

00409 2

Second Edition, First Printing, 1983

Published by Arco Publishing, Inc.
215 Park Avenue South, New York, N.Y. 10003

Library of Congress Cataloging in Publication Data

Ballantyne, Verne, 1904–
  How and where to find gold.
  Bibliography: p. 143
  Includes index.
  1. Prospecting.   2. Gold.   I. Title.
TN271.G6B34   1982   622'.1841   82-13734
ISBN 0-668-05377-1
ISBN 0-668-05385-2 (pbk.)

Printed in the United States of America

And the gold of that land
is good: there is bdellium
and the onyx stone.

Genesis 2:12

Acknowledgments

My good friend and associate, Bill Little, was the important link between me and the old time prospectors and miners. The information given him by the early prospectors is the basis for this book. The personal incidents related are his. Bill started his career in mining shortly after the end of World War II when he took up residence in La Porte, California. The name of that little town tells a lot right there. If you look it up on the map you will see it is right in the heart of the gold country in Plumas County in north central California.

Charles Scott Haley, one of California's most renowned geologists and mining engineers stated in his historic Bulletin No. 92, Gold Placers in California, "The principle gold placer area of California lies in the Sierra Nevadas between Susanville on the north and Mariposa on the south."

The historic old mining town of Quincy is the county seat and La Porte is 30 miles southwest over one of the most crooked, roughest mountain roads you can imagine. Strawberry Valley, Diamond Springs, and Poverty Flat aren't far away. Downieville and Johnsville are a little farther and Reno, Nevada is about 125 miles as the crow flies, southeast of La Porte, over the main range at an altitude of something over a mile. The snow gets 10 to 15 feet deep up there in the winter and if you don't plan to stay in for the season you better get out before the first big snow comes in November.

The south fork of the Feather River is not far from La Porte and there are a good many well-known old deep channel placer mines not far away. Among these are the Red Ravine, the Feather's Fork, and Bellevue Mines. Bill spent over 30 years in and around the La Porte area, and when he first landed there a good many of the real old-timers miners and prospectors were still around. He absorbed a lot of the lore of the early gold mining days and a wealth of information on prospecting and mining which the early miners had learned from their experiences and what had been handed down to them by the old-time miners ahead of them.

Later Bill staked his own claim and learned more about mining by his own personal experience. One of the incidents he encountered had to do with the handling of dynamite—blasting—which is recorded in the chapter of this book entitled "Powder—Fire in the Hole!" He had a narrow escape that taught him some safety measures we have been able to pass on to you. He was not involved in a later incident which resulted in the death of several people but was so concerned about it that he analyzed the cause of the accident and developed a method of handling the electrical system of detonation of dynamite that would have prevented the accident. This valuable information is part of this chapter.

For several years Bill owned and ran a core drilling rig and did custom core drilling work for miners in California and Nevada. The picture in this book of the core drill is one he built and operated.

Bill has his claim out from La Porte and does his assessment work each summer hoping that some day he will be able to sink a shaft down to what he believes is a rich deep channel placer deposit. Several years ago he moved to Arizona where he lives in a small mining town. He has done a lot of prospecting there and staked a dozen claims on a faulted area that has one of the widest veins of the most beautiful quartz you can imagine. It is wider than the Mother Lode quartz vein in California. Here is hoping Bill's long experience in mining pays off real big.

The information for the new chapter, How to Build and Use Your Own Mini-Rocker, was provided by Bill Hinsen. Bill is a native Montanan that I have known since we lived on nearby farms back in 1920. Bill is a natural outdoorsman. As a small boy he was a student of nature and wildlife and ran his own trap line before he graduated from grade school. Later, he held down a full-time job at the local flour mill for a good share of his life but always managed to keep his trap line going during the trapping season.

Bill's outdoor instincts, along with trapping, led him into prospecting. During the past 25 years he has become somewhat of an expert at panning for gold. He has taught classes in panning and ran a panning demonstration, a tourist attraction in the historic mining town of Virginia City, Montana for several seasons. He has an inventive turn of mind and 13 years ago decided to build a miniature version of a gold rocker to be used as a prospecting tool. He worked with different models and versions for several years before he developed one that satisfied the objectives he had in mind, which was a lightweight tool, one easy to build, efficient in its operation and simple to use.

From his experience in teaching gold panning it was his opinion the ordinary person does not become a safe panner in a few easy lessons. He believes it takes a great deal of practice to be able to pan out a sample of gravel and save a high percent of the values. "Most beginners lose over half of the gold in their pan," he says. With his carefully designed "mini-rocker," as he calls it, the first-time novice prospector will save practically all the gold. "If there is gold in the sample it will be in the riffles of the rocker," he says. "There is not any other place for it to go. If there is no gold in the riffles there was none in the sample, it is just that simple."

We thought there should be a chapter in the book to give a beginning prospector the complete plans on how to build a mini-rocker for himself and how to use it for better, more accurate prospecting. Together we developed the chapter and hope it will encourage you to build your own rocker and get out in the gold country and find some gold. The plans were prepared by an assistant professor in the Agricultural Engineering Department at Montana State University, Lee Erickson, Ph.D.

Terry Groth, a local young man in the sign business in Bozeman, Montana caught the spirit of the text and by his creative talent produced the art work which has added so much to the book. It has been well said that one picture can tell more than a thousand words. In my opinion, Terry's sketches tell even more.

The Montana Bureau of Mines and Geology at Butte graciously gave permission to use their working drawings of the sluice blox and cradle. In addition, their evaluation of the pan as a prospector's tool and as a device for actual gold recovery was most helpful.

The California Division of Mines and Geology gave valuable help by authorizing the reproduction of a typical gold diving operation from their April 1972 issue of California Geology.

The new material on federal regulations was provided by the personnel at the supervisor's office of the Gallatin National Forest of the United States Forest Service, among whom were Betty Smith, Sherm Solid, Mike Williams and John McCulloch.

Contents

Preface to the First Edition

There is still plenty of gold to be found in the West! Contrary to what many people think, all the gold in the United States was *not* mined out by the old '49ers or by the Depression Miners of the 1930's. This may have been true of the famed gold bearing creeks of the Mother Lode Country of northern California, but many lesser known areas were not worked as extensively. There are miles of buried gold placers in ancient streambeds in the High Sierra Mountains of California and other places that have never been touched. Many geologists believe that there is many times more gold in the ground than has ever been mined!

There is gold to be found in many of the tributaries of the Snake, Columbia and Salmon Rivers in Washington, Oregon, and Idaho. The Bitter Root, the Clark's Fork, and the North and South Forks of the Flathead River in Montana are all part of that great river system. The Kootenai, a little known but scenic and dashing river that loops down from the new gold fields of British Columbia into northwestern Montana and the Panhandle Country of Idaho, also offers good possibilities.

The High Country of south central Montana is an empire of its own. It is traversed by a network of tributaries of the Madison, Jefferson, and Gallatin Rivers which meet near Three Forks, Montana to form the mighty Missouri. The old gold mining towns of Virginia City, Alder, and Nevada City on Alder Creek, to name only a few, testify to the presence of gold in the early days, and who is to say it was all mined out?

Also, the countless mountain streams in the Rocky Mountains of Western Canada, the Northwestern Territories, and Alaska are all part of the Gold Belt. Much of this is virgin country waiting for modern prospectors to seek their fortunes.

Consider this: gold mining did not die out because there was no more gold! The demand for labor and material during two World Wars put economic pressure on gold mining. The wages offered to boost industrial output pulled men from the gold fields, and then Government restrictions clamped down on non-strategic gold mining.

Following World War I, the fixed price of gold at $35 an ounce could not cover the cost of production. It was impossible for most of the hard rock mines to reopen, nor was there any incentive for placer miners to return to their diggings. Gold mining could not compete with the higher wages paid in other segments of the economy.

Between 1972 and 1976, the official price offered by the Treasury of the United States of $35 (and later $42) an ounce was challenged by a free world market price of $150 to $200 an ounce. Further increases are likely. Consequently, a new interest in gold mining is growing proportionately.

In 1974, deteriorating confidence in so-called lawful (paper) money, with no backing in gold or silver and very little purchasing power, prompted the 93rd Congress to repeal the infamous Gold Reserve Act of 1934 which had made private ownership of gold illegal. Thus a right which had been denied United States citizens for over 40 years was again restored.

The effect of this Act on gold mining in the United States, coupled with the phenomenal rise in the price of gold on world markets, can only be guessed. Could it be that the Gold Rush of the 1970's is just beginning?

Verne H. Ballantyne
Bozeman, Montana

Preface to the Second Edition

The reception given to the first edition of this book, HOW AND WHERE TO FIND GOLD, which was published in 1976, has been most gratifying. The favorable comments from amateur prospectors who have found the information helpful to them in their search for gold has been pleasant indeed. I have been especially grateful for the favorable remarks from the more experienced prospectors who have in effect given their stamp of approval on the book.

A lot of water has gone under the bridge since the first edition was published. As anticipated, there has been a great deal of interest in prospecting during the past five years. A brief glance at the price action of gold gives a graphic picture of the explosive interest in the yellow metal during this time. From the high of $200 an ounce in 1978, the price of gold moved up gradually into the upper $300 level in the latter part of 1979. Then the price leaped to the unheard of figure of $877 by January of 1980. During the following three months the market collapsed to a low of $450 in March then rebounded to $725 in September—all in the same year. The failure of the price to even come close to the high of $877 marked the beginning of a down trend which is still continuing at this writing in late 1981. You will have at least a partial answer to what happened next when you read your current market news. It is a continuing story.

Prospecting for gold until this phenomenal rise in price had practically become a lost art. The old-time prospectors had long since disappeared from the scene and few young people had taken their place. The price action stimulated a tremendous demand for information on prospecting methods and a heavy demand for the first edition of the book. The publishers found themselves scraping the bottom of the barrel on the second printing in late 1981 as the supply became exhausted.

In view of important, new federal regulations relating to prospecting since 1976 and the new gold prospecting tool, the mini-rocker, it was decided a revision was needed to include this new information. Since the basic information of how to look for gold and where it is to be found remains the same, little change has been made in those parts of the book.

The new chapter, "How to Build and Use your own Mini-Rocker," gives a list of materials, working drawings and step-by-step instructions on how to build and use the mini-rocker. This was first prepared as a special study and published in a limited quantity. It has been well received and is being used in junior and senior high schools and city libraries as resource material. It is a popular booklet and is sold at the gold shows sponsored by the Gold Prospector's Association of America in the western states each year.

The chapter entitled "Ownership and Marketing" has been expanded to include much new information and the new chapter, "Federal Regulations" includes the recent requirements on the protection and preservation of the surface of the public lands and the proper recording of unpatented mining claims. The information is presented in easy question-and-answer form.

The gold prospectors of the 1980's have opportunities and advantages never before available. Old methods are being revised or supplanted by new technology and modern machines not dreamed of in previous times. Known and available to all are the areas where gold can be found. A wealth of geological information is free for the asking from mining bureaus in each state and province where there are mineral resources. Geologists and mining engineers employed by the states are available to any prospector or small miner.

Geologists estimate that not over five to ten percent of the world's gold has been found and mined. The '49ers of a century and a half ago found millions and the depression miners of the 1930's found twice as much, with much less publicity. But together they only scratched the surface. The rest is out there waiting for the modern prospectors of the 1980's.

The purpose of this book is to provide practical, easily understood and useful information for the new or beginning gold prospector, not only of how and where to find gold, but how to stake the claim, produce and process the find. He or she is also given a brief look at mining hard rock gold, how to blast with safety and how to live with government regulations. While this book is written primarily for the person who has had no previous experience or knowledge of prospecting or mining, it is hoped that some who have tried their hand in it before may find information of value to them as well.

V. H. B.

Introduction

In American folklore, a bewhiskered old man leading a burro symbolizes the search for gold. The prospectors who found the prized yellow metal in the deserts and mountains of the West did so with pick, shovel, gold pan, and sweat. They survived on bacon, beans, coffee, and grim determination. Hard, patient work and trial and error developed their knowledge of how and where to find gold. This precious knowledge, born of experience and only shared with other old timers, produced a reservoir of secrets not found in any textbook. The campfires where tall tales and true adventures were told were the prospector's trade schools. Wherever gold prospectors met, the talk was about how and where to find gold. Although the old gold mining days are history, the secrets of the old prospectors have lived on. This book will tell you about these colorful men and the mining skills they used in the days when gold mining was a way of life and a crucial economic venture.

You may be wondering how I came to have the gold-finding knowledge of the old timers. That is an interesting story in itself. In the mid-1900's, some of the old timers were still around and actively seeking gold. As a young man just out of the Air Force, I was attracted to the old-time prospectors and the sourdough. I have sat by the campfire many a night and re-lived through these men their experiences on the trail of gold, and I have listened to their theories about how Nature made gold and hid it away for men to find in due time and season.

These men were eager to pass on their secrets. They let me work

Gold mines were not closed because there was no more gold! They were closed because the Government's fixed price of $35 per ounce would not permit the miners to cover the cost of production.

Prospector's seminar.

with them on their claims and taught me how to work a digging. I became an amateur member of their circle and eventually staked my first claim. That was over twenty-five years ago. The search for the yellow metal has been a prime interest in my life, along with an interest in related mining operations. The wisdom of the early prospectors has been enhanced by modern technology. The burro or pack horse has been replaced by the four wheel drive pickup and camper. But nothing can replace the gold pan and pick, even though a gas-engine powered suction pump and portable sluice box come in handy. The modern miner still has a lot to learn from the old prospectors about where to look for gold and how to find and recover it. In this book, we have combined the best of the old secrets with newer methods and equipment.

My hat is off to you, partner, and may your prospecting days be many and your rewards be great.

William A. Little
LaPorte, California
1976

Chapter I
Where to Look for Gold

Until the old '49ers found gold in California, no one except the Indians and the Spaniards knew it was there. It is common knowledge now that there is gold in all of the 11 western states, Alaska and Canada. That doesn't mean you can find gold just anywhere in those states, or even in specific areas of those states, but it does help to know that there are certain parts of the country where gold definitely exists. (See page 2.) The first rule to guide you in your search for gold is:

Look in places where gold has been found in the past.

You will probably do your looking for gold close to where you live, as this is likely to be a weekend or part-time venture. Write to the Bureau of Mines in your state and request information on the location of known gold fields. If you are seriously considering looking for gold on a full time basis, write to the United States Bureau of Mines in Washington, D. C., and request maps and information on known gold producing areas in the United States and Canada. You might consider leaving home in this case—but think twice before you do. Prospecting is a highly speculative activity even for an experienced person. It carries no guarantee of providing a living.

Placer Gold

Placer is a word of Spanish origin defined as "an alluvial or glacial

1

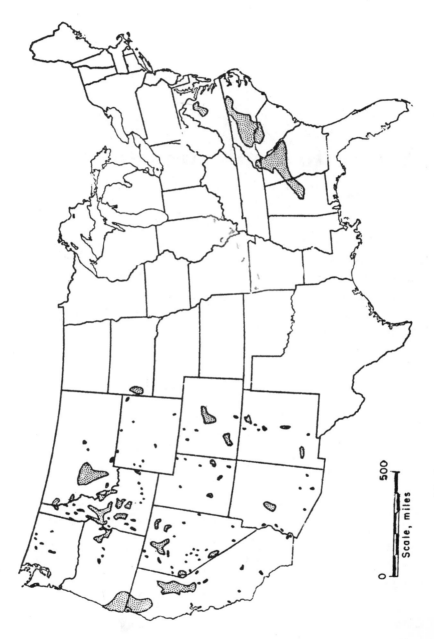

Gold placer areas of the United States (Courtesy U.S. Department of the Interior, Bureau of Mines, Information Circular, How to Mine and Prospect for Placer Gold).

The Gold Belt extends from the upper part of North America (Alaska), through Western Canada, the United States, Mexico, and Central America, continuing along the Western portion of South America. It is associated with the chain of mountains traversing the Western portion of the two continents.

deposit containing particles of gold or other valuable mineral." Undoubtedly, the first discoveries in North America by the Indians of the attractive yellow metal were from placer deposits. These were later worked by the Spanish.

The historic discovery in California at Sutter's mill in 1849 in what is now El Dorado County was a placer discovery and most of the early gold rush was for the gold found in the rivers and streams of California, Southern Oregon, and other western states.

Quartz—A Source of Gold

The old timers agreed that the free gold found in the streambeds and bars in the form of nuggets, flakes, and dust had to come from an original source, and this source was usually quartz. For some reason, unknown to the old timers or to modern geologists, gold is almost always found in or very near quartz rock.

Weathering, upheavals, and erosion broke up the quartz and gradually released the gold from the veins or streaks in the rock. The released gold was then subjected to the forces of erosion, and the grinding or polishing action of rock and water. Small pieces of gold were literally ground to dust, and larger pieces were reduced to small, rounded nuggets. This was the source or origin of the free gold known as placer gold.

Gold Is Heavy

Aside from its yellow color, the most distinguishing characteristic of gold is its weight. It is 19.7 times heavier than water and at least seven times heavier than any rock with which it is found. The weight of gold allows us to separate it from other mineral material. This factor is basic in most of the equipment used in the recovery of gold, as we shall see later.

Hard Rock Gold

The gold found in place in the quartz rock is called vein or hard rock gold and its recovery requires an entirely different procedure than the methods used in recovering placer gold, which will be discussed later.

Bedrock

Bedrock is the solid rock mass underlying the superficial formations and soil. Knowing about bedrock is important to us because a large proportion of placer gold is found on top of bedrock. Because of its extreme weight, gold usually washes to the bottom of the surface material. Not all gold finds its way to bedrock but that is where the larger nuggets and flakes are usually found. The illustration on page 6 shows likely locations on the bedrock where gold may be lodged.

Potholes, Riffles, and Rapids

The inside bend of the creek where large boulders or outcroppings are present is a likely place to find gold. Gold tends to lodge behind or on the downstream side of these boulders, in pockets or potholes formed by the current of the stream. Collect your gravel samples at varying distances from the boulder, as close to the bedrock as possible. Another promising place is just above any natural shelf which crosses the stream bed. Shelves are indicated by riffles in the stream where the water slows to form a pool where gold may collect. The heavier gold will settle out first and sink to the bottom of the streambed. Natural breakwaters created by the bank or by a fallen log may also provide places where the running stream will drop its gold. These areas are called creek placers to distinguish them from other types and places of gold accumulation, and they are considered to be the most important location for placer gold.

Bench or High Bar Placers

Another important source of gold is the area immediately above the present creek bed. Along these banks there may be gravel deposits or bars on top of the bedrock where the gold has been deposited along what was formerly the wider creek bed.

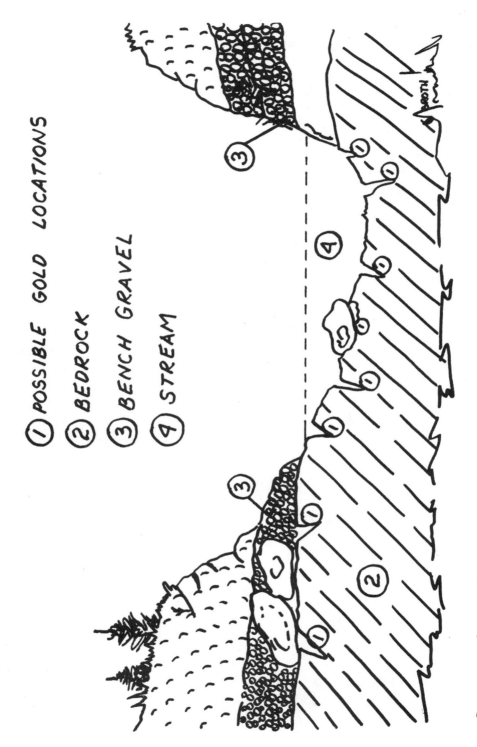

POSSIBLE GOLD LOCATIONS

① POSSIBLE GOLD LOCATIONS
② BEDROCK
③ BENCH GRAVEL
④ STREAM

Cross-section of a stream bed.

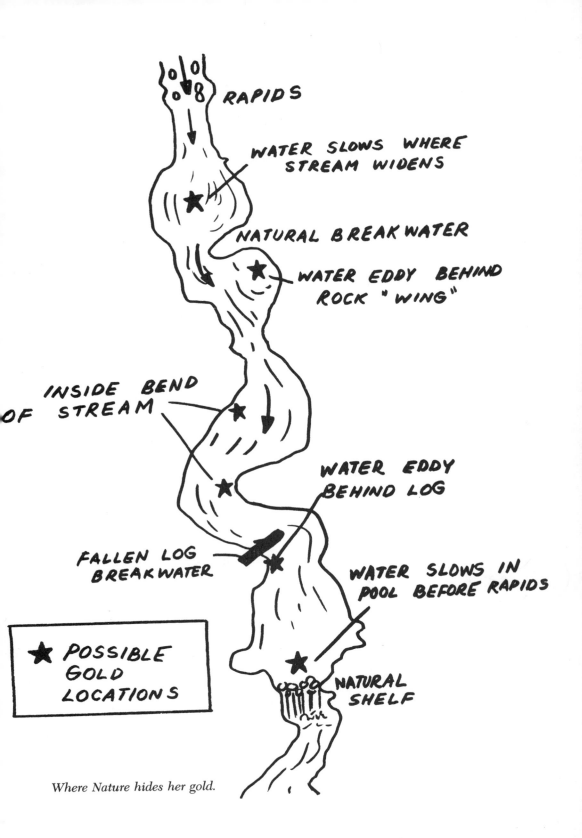

RAPIDS

WATER SLOWS WHERE STREAM WIDENS

NATURAL BREAKWATER

WATER EDDY BEHIND ROCK "WING"

INSIDE BEND OF STREAM

WATER EDDY BEHIND LOG

FALLEN LOG BREAKWATER

WATER SLOWS IN POOL BEFORE RAPIDS

★ POSSIBLE GOLD LOCATIONS

NATURAL SHELF

Where Nature hides her gold.

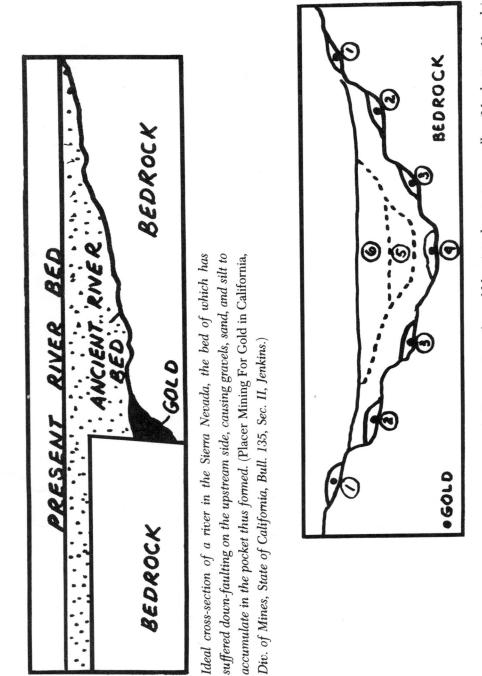

Ideal cross-section of a river in the Sierra Nevada, the bed of which has suffered down-faulting on the upstream side, causing gravels, sand, and silt to accumulate in the pocket thus formed. (Placer Mining For Gold in California, Div. of Mines, State of California, Bull. 135, Sec. II, Jenkins.)

Cross-section of a gold-bearing desert stream valley (Manhattan, Nevada), showing the results of several periods of stream deposition from the oldest (1) to the youngest (6). (After Ferguson, U.S. Geological Survey Bull. 723.)

Crevice Mining

The beginning prospector can start his search for gold in several ways and with varying amounts of equipment. One approach that takes a minimum of equipment and will get you started in this most interesting and challenging activity is to do some crevice mining. Crevice mining is a specialized type of placer mining. The name comes from the fact that one of the most likely places to find gold in a mountain stream is in the crevices of the bedrock over which the stream flows. Gold has accumulated in these cracks and crevices over a long period of time.

The old timers found that gold was located in the cracks and crevices in the bedrock not only in the present streambed but also higher up the bank and in exposed bedrock that the stream had reached during spring runoff. Gold was hidden in dry streambeds as well. Some of the early miners made a special search of these cracks and crevices. They were called crevice men or crevice miners. Crevice men who didn't bother to stake claims and ranged up and down the creek working on any ground not staked were called snipers.

Tools Required

Crevice work is probably the easiest way to start looking for gold. To be a crevice miner, you need specialized tools to get down into those cracks in the bedrock. I'm not going to send you to the store to get a lot of expensive equipment. You probably have some of the tools you'll need at home. Here is a list of the crevice man's tools:

1. Large tablespoon.
2. Long handled teaspoon (tall glass variety).
3. Old screwdriver, bent two to three inches from the end of the blade. Makes an ideal digger.
4. Tweezers. These are for picking up flakes or small nuggets out of the sand in your pan.
5. Small glass jar with a tight fitting screw cap with a wide mouth. Put your gold in this.

6. Small magnet. This is useful in sorting out the heavy magnetic pieces of black sand after your panned sample has dried.
7. Small shovel with short or long handle, whichever you prefer.
8. Gardener's hand scoop.
9. Miner's hand pick.
10. Magnifying glass. This will add interest and make identification easier. The folding, double lens type is preferred.
11. Flashlight.
12. Galvanized metal bucket. It has many uses and is handy for carrying your small items to the site of your diggings.
13. Gold pan.

The last and absolutely indispensable item is a gold pan. Some of the old timers may have used a frying pan in a pinch, but I am sure they got a regular gold pan at their first opportunity. It is designed and shaped for panning gold. Several sizes are available, but the size generally referred to as standard is 16 inches in diameter at the top and 10 inches at the bottom, with a depth of $2\frac{1}{2}$ inches. I recommend the half size pan which has a top diameter of 12 inches, a $7\frac{1}{2}$ inch bottom, and a depth of 2 inches. Level full, this pan weighs 9 pounds—as compared to 20 for the 16 inch pan. It is easier to carry and easier to use in a small stream or tub. Its lighter weight when filled is less tiring. An experienced panner can wash two of these half sized pans of gravel in less time than it takes to wash one standard size pan.

A new pan is usually coated with some type of oil, grease, or other rust preventive. This must be removed before using the pan, as even the slightest trace of any oily substance (even body oils which may accumulate from your hands in normal use) will cause fine gold particles and dust to ride out over the edge when panning. You must clean your pan before using it. The surest and easiest way to clean your pan of this oil is to pass it back and forth over a gas stove burner or similar flame until the metal turns blue. Even though this is called burning or blueing the pan, care must be taken not to heat it unevenly or excessively. Excessive heat under one spot can warp your pan and make it useless. Blueing the pan not only does away with the grease, but also provides a dark blue color which is an ideal contrast to fine gold particles.

Crevice miner.

Crevice miner's tools.

Gold pan

Pail

Bent file

Tablespoon

Long-handled teaspoon

Magnifying glass

Tweezers

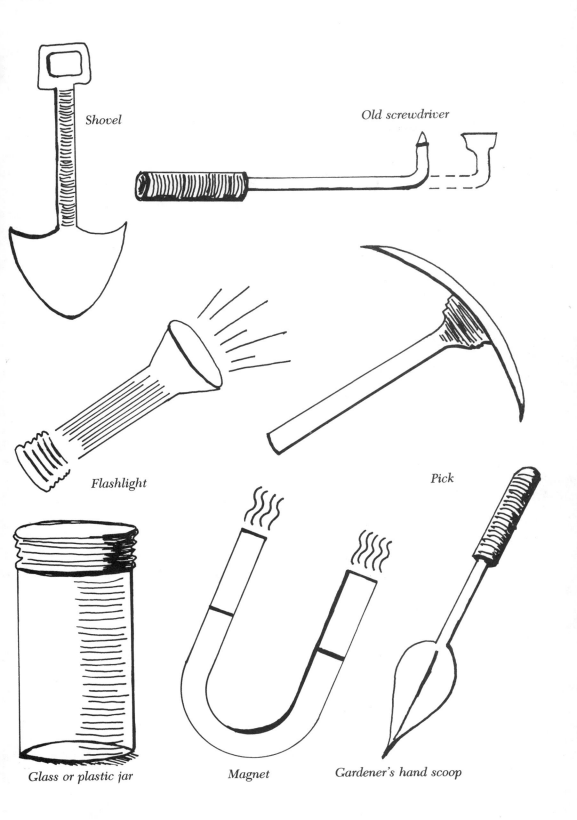

Shovel

Old screwdriver

Flashlight

Pick

Glass or plastic jar

Magnet

Gardener's hand scoop

Mining the Crevice

Mining the cracks, crevices, and fissures in the bedrock is our present objective. Pick out your crevice. It is probably filled with silt, sand, small gravel, and vegetable matter. Using the bent screwdriver, loosen this material so that you can scoop it out of the crevice with your spoon and dump it in your gold pan. Dig all the way to the bottom of the crevice. Remember that the gold, answering the call of gravity, has probably been working its way to the bottom of that crevice for a long time. You may have to enlarge the crevice or open it up a little with your pick. Go ahead and dig down into the crevice! This is the way to learn on your own.

Fill your pan two-thirds full. Include the plants growing in the crevice. Their roots will have entered the very narrowest bottom part of the crack and the tiny root hairs may have grasped a flake or small particle of gold. If so, it will be released when you shake the roots around in the water in your pan.

Now you're ready to pan your sample!

Chapter II
Panning for Gold

You have a pan and are ready to take a first sample. Follow these steps until they become natural and automatic:

1. Preparation

Having filled the pan two-thirds full of sample material from the crevice, choose a shallow place along the bank of the creek that is a little deeper than your pan. The stream should be moving fast enough to keep the water clear, but slow enough so that it won't wash any of your sample away. (A tub of water will do if you aren't near a creek.) Carefully submerge your pan until it is resting on the bottom of the creek. When the material is thoroughly wet, work and stir the mixture around with your hands until any lumps of dirt are disintegrated. Break any hard lumps between your thumb and fingers. Dirt and sand adhering to any plant roots should be washed into the pan. Then throw out the plant material. Any clay present must be stirred up until it is entirely dissolved and washed away. The large gravel and rock can be picked out and discarded.

2. Suspension and Stratification

Next, grasp the still submerged, level pan by its opposite sides and perform a vigorous left-right, left-right motion. This, plus a clockwise,

One of the last of the old time prospectors demonstrates the fine art of panning for gold. Taken in the summer of 1955.

counter-clockwise motion, should keep the contents loose but not agitated. This permits the separation of heavy materials from lighter ones. The heavier minerals will sink to the bottom while the lighter gravel and rock material will be displaced upward. Most of the larger gravel can then be raked over the edge of the pan with your fingers.

3. Washing

Washing is accomplished by slightly raising the near side of the pan above the water, leaving the far side submerged. Switch from the right-left, right-left motion to a circular, swirling one. This will gradually work the lighter material over the submerged side of the pan into the water. The edge of the pan can be raised or lowered to regulate the washing. A side-to-side motion is also useful at this point, used alternately with the circular action.

4. Cleaning

The objective of this step is to wash away the remaining sand and pebbles from the heavier material on the bottom of the pan. The pan is jerked vigorously from side to side to loosen the material and cause further stratification. At the same time, the pan is tipped forward gradually, until the sand bed is flush with the pan's edge. Stop briefly at this point to allow the bed to settle.

Your pan is now at an angle, with the lower edge barely under water. Maintaining this angle, dip the pan vertically into the water four or five times, so that the bed is slightly covered. As you raise the pan, you will see lighter particles wash out over the edge into the water. When this lighter material is washed away, the pan is again partially filled with water, and the bed loosened and restratified by repeating the vigorous side-to-side motion. Repeat the dipping, washing, and shaking until only the heavy mineral concentrate remains in the pan. This may seem awkward or tedious at first, but you will soon develop your own variations on these basic techniques.

5. Inspection

The panning operation is complete when the original sample of material has been reduced to a small quantity of black sand and minerals. Inspection of the sample is made by placing a small amount of water in the pan and swirling it gently, so that the water moves the lighter particles ahead of the heavier ones. The gold is brought into view at the end of the wash of material, if you have been working in gold country.

6. Recovery

Larger pieces of gold can be picked out with your tweezers and dropped into your glass jar. Smaller pieces and flakes will adhere to the end of a wooden match long enough to be put into the jar. If there is a considerable amount of fine, black sand, the sample should be dried and the sand removed by placing it on a dry pan, or a stiff piece of paper or cardboard. When the sand is dry, blow gently across the surface while tapping the pan or paper. The sand will quickly blow away from the gold and any heavy magnetic material. Your pocket magnet can then pick up the magnetic material and the gold can be scooped up with a stiff piece of paper and poured into the flask. If the magnet is wrapped in plastic, the material on it can be discarded easily when the magnet is removed from the plastic.

1. Preparation—Mix and stir to separate all the parts of the sample.

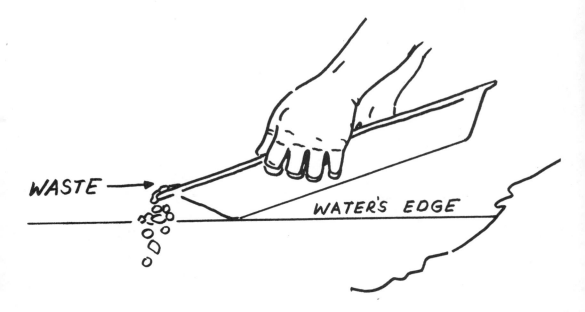

2. *Suspension and stratification—Give the pan sharp jerks clockwise and counter-clockwise. Rake large waste over the side of the pan.*

WASTE →

WATER'S EDGE

3. *Washing—Tip the pan and give the pan a circular motion to dispose of waste material.*

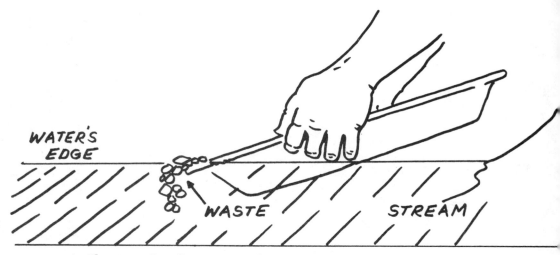

4. Cleaning—Dip the pan several times up and down, and vigorously from side to side.

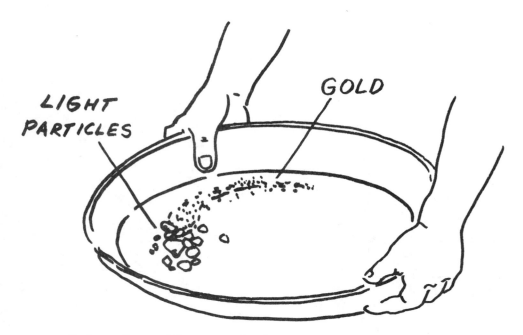

5. Inspection and Recovery—Swirl contents around with a small amount of water to move light particles ahead and expose colors. Remove gold with tweezers.

Mercury

Mercury is often used to separate the fine gold particles and dust from the black sand. When the sand and gold mixture is thoroughly dry, pour a little mercury into the pan and roll it around until it picks up all the gold. To recover most of the free mercury, pour the gold-mercury amalgam onto a piece of chamois skin or fine canvas, then fold the four corners over to form a pocket containing the amalgam and squeeze as much of the mercury as possible through the skin or canvas into a dish. The remaining ball of amalgam can be reduced to its separate components with this trick used by the old timers. Place the amalgam in a hole scooped out of the center of a raw potato. Then place a tin plate over the hole, turn all of it over, and heat it on a stove until the potato is partly cooked. The gold will be found on the plate in the form of a sponge. The heat will vaporize the mercury (as it has a very low boiling point), and it will condense and remain in the cooler flesh of the potato. The mercury can be recovered by placing the potato in a dish of water. The mercury will sink to the bottom. *EXTREME CAUTION must be taken so that the mercury vapor does not escape into the air. MERCURY VAPOR IS VERY POISONOUS.*

Grizzley Pan

A trick that will speed up the panning operation immensely is the use of a sieve called a grizzley pan. It is made by drilling a number of quarter inch holes in the bottom of a pan the same shape and size as the one used for panning. Place the grizzley inside the regular pan, fill it with your pay gravel, and submerge it in the water in the usual manner. When the mixture is well soaked, lift the grizzley slightly and rotate it sharply back and forth under water until all the material less than a quarter inch in diameter has passed through into the regular pan. The larger material can be checked quickly for any large nuggets and then can be discarded. The fine material in the regular pan is washed in the usual way.

Safety Pan

Occasionally, even an experienced panner will accidentally spill his carefully washed sample into the creek. To prevent the use of profanity and to save what might be a valuable sample, it is a good practice to direct the pan tailings into a second pan which is called a "safety pan." You can decide for yourself if the use of the safety pan is worth the added bother.

Practice Panning

Successful panning is a skill that is acquired largely by practice and experience after a certain amount of instruction. We suggest that you practice at home before you head off to seek gold. Conditions very similar to those encountered in placer mining can be simulated by using ordinary sand and gravel, your gold pan, a tub of water, and fifty or so buckshot or very small fishing sinkers. You can even cut some of the shot up to simulate fine particles of gold. This isn't very exciting or romantic, but it *is* practical. With your gold pan and these materials, you can follow the instructions and test your proficiency. Fill your gold pan two-thirds full of sand and gravel and then add the fifty buckshot. If you get to the bottom of your pan and find only half of the shot left you are doing well for a beginner, but you can do better. Suppose those other 25 buckshot had been gold nuggets that you washed into the creek! Practice until you find 45 to 48 shot left in your pan. Then you will know that you are good enough to get the most from your samples.

Fool's Gold

One particular bit of mining folklore had to do with "fool's gold," the common name given to iron pyrite. Chemically speaking, iron pyrite is a combination of iron and sulfur. It is yellow and it frequently appeared in the bottom of the gold pan. It was called fool's gold because inexperienced miners mistook it for the real thing. It can be easily distinguished from gold, however, as it is easily crushed. Gold is highly malleable; it will not crush or break up. Iron pyrite is deceptive. Oddly, gold is often found near it, sometimes in the same rock

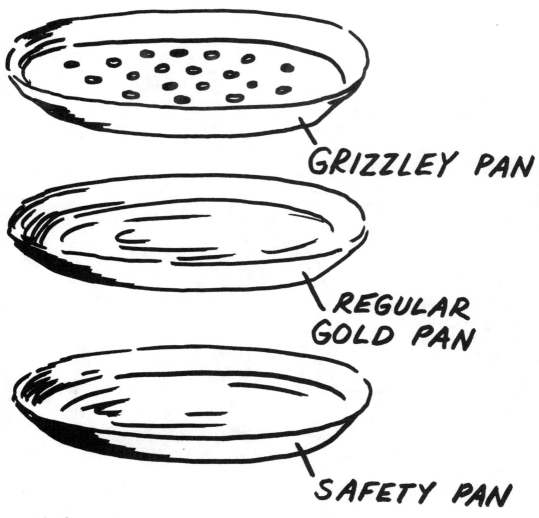

GRIZZLEY PAN

REGULAR GOLD PAN

SAFETY PAN

The three pans.

formation. For this reason, the old prospectors did not ignore fool's gold.

Evaluation of Panning

To put this whole matter of panning in perspective, a few observations can be made. Panning is an indispensable way for the placer miner to analyze the gold content of a prospective diggings that may then be worked by other, more rapid and productive means such

Practice panning.

as the cradle or sluice box. The gold pan alone has recovered a good amount of gold for old-time miners, Depression miners, hobby and Sunday afternoon miners and other amateurs, but it does have its limitations from an economic point of view.

The Economics of Panning

According to the estimates of the Montana Bureau of Mines and Geology in their Miscellaneous Contribution No. 13, an experienced man working with a standard size gold pan will do well to pan sixty pans in a ten hour day. At twenty pounds per pan, he will handle 1200 pounds of gravel. At this rate, it will take him three days to pan one cubic yard. If the gravel produces $15 worth of gold per cubic yard, which is fairly good gravel at a price of even $100 per ounce of gold, it is easy to see that it would take an entire day to recover five dollars worth of gold. The average for the Depression miner looking for gold selling for $35.00 per ounce was more like one dollar per day. Consequently, panning gold under average conditions is not a bonanza. It is a prospecting tool and should not be considered a good

Frank Riley, foreman of the Feather's Fork deep channel placer mine when it was closed down in 1934, washes some tailings from the old dump in August 1974.

method of serious or commercial gold recovery, except under exceptional circumstances.

Panning for Fun

Of course, not everything is done strictly for economic reward. From the point of view of a hobbyist, weekend camper, or prospector, the reward of panning for placer gold may come largely from being in a natural and perhaps scenic environment where the air is fresh and the skies are blue. Any additional reward in the form of gold recovered is an extra benefit.

Chapter III
How to Build and Use Your
Own Mini-Rocker

The mini-rocker is an adaptation of the old, tried and proven, standard gold rocker or cradle which was developed to a high state of perfection during the '49 gold rush days. The original idea was developed in the south, perhaps Georgia or Alabama, many years before the California gold rush. At first it was just a hollowed-out log with one long riffle in the bottom and a handle on one end to rock it.

Some of the experienced prospectors from the south brought the old log washer idea with them when they joined the California gold rush. It soon developed into the California gold washer with its hopper, baffle and riffle bed and later became known all over the gold country as simply a rocker or cradle. It was a production tool. It was, and still is, one of the most efficient methods ever developed for separating the gold and black sands from the gravel.

Not until Bill Hinsen developed the miniature version was it ever thought of, or used as, a prospecting tool in the same sense as the prospector's gold pan.

The mini-rocker has certain advantages over the romantic and historic gold pan as will be explained in the text of this chapter.

Before going into the details of how to build and use your own mini-rocker I want to tell you some other things about it. To begin with, the mini-rocker is a brand new prospector's tool. It will test your

The mini-rocker is a simple but highly effective tool for the prospector. It is fast, accurate, and positive in accomplishing the purpose for which it was designed—to test your sample and produce gold.

sample in a fraction of the time it takes with a pan, and it will produce gold if there is any in the sample. You can save money by building your own.

Comparison—Mini-rocker vs. Pan

Most people automatically think of a gold pan and panning when there is a discussion on prospecting for gold. And for good reason: It was the only tool and method available before William J. Hinsen developed the mini-rocker. In spite of all the romance and history associated with the gold pan and panning for gold, let's face some facts. Except for the very experienced person, it is inaccurate and slow. In addition, it is just plain hard work. Crouching down, in a strained, awkward position, with your knees on a hard rock and your hands in the icy cold water of a high mountain stream, is not much fun. And I'll wager the beginning prospector will loose at least half the values in the sample over the edge of the pan. It takes a lot of practice and experience to be able to pan out a sample of gravel and save a high percentage of the gold. It will take a beginning prospector at least five minutes to pan a sample, more likely, 15 minutes to a half hour.

Comfortable—Fast—Accurate

But, the mini-rocker has changed all that. With it you can sit on a flat rock or stool, in a comfortable position, while you test your sample, and there is no need or reason to have your hands in the water at all. After reading the instructions in this book on how to use your mini-rocker, and after a few minutes of practice, you will be able to test a shovel full of gravel in less than two minutes. And the best part of it is that you will know the results are accurate. If there was any gold in the sample it will be found on the mat on the bed of the rocker. If there are no colors, then you will know there was no gold in the sample. It is as simple as that. With a gold pan you always have an uneasy feeling that some of the gold may have slipped over the edge back into the water. You never know for sure.

Development

The mini-rocker is a small version of the rocker or cradle used so successfully by the old-time prospectors to produce gold in the early gold rush days. Bill developed it to fill a need, shared by all gold prospectors, for a light, easily carried tool that would quickly and positively, *test* for gravel for gold. If found, it would *produce* enough on the spot to make it worthwhile. When a man goes fishing he isn't satisfied to just find where the fish are; he wants to bring some fish home. With your mini-rocker you will bring home some gold if it is in your samples.

Bill has been a prospector, miner and trapper all his life in the Rocky Mountain country of southwestern Montana. After two years of trial and error development work he hit on just the right size for the mini-rocker. It is small enough to be easily carried, and large enough to produce some gold to bring home for evidence as well as find it. Since them, it has been in use in the gold country of the western states and Alaska in a limited way for the past eight years with great success. Until recently only custom-built units Bill produced were available.

Here's How it Works

The mini-rocker, like the full-sized rocker used so effectively by the old time miners, consists of a hopper and trough. The hopper is built separately so it can be slipped inside the trough for compact, easy carrying. It has a ½-inch screen in the bottom, beneath which is a trough with riffles to catch the gold. The hopper is filled with gravel and water is poured into it. The device is rocked back and forth and the gravel which cannot pass through the screen is thrown out. The gravel containing the gold passes through the screen into the trough. The gold, being much heavier than the sand and gravel, immediately drops down in front of the riffles and is retained by them while the lighter sand and gravel passes on over the riffles and out the end of the trough.

The rolling of the gravel and the swirling, cross-currents of the water, generated by the action of the mini-rocker, creates an ideal condition for the separation of the gold from the sand and gravel.

There is no chance for the gold to be lost. If there is gold in the gravel, it will be found in the first six inches of the mat under the baffle. If there is no gold in the rocker bed, you know there was no gold in the gravel.

Saves Gold and Time in Comfort

Anyone who has panned for gold knows it takes a great deal of experience to be sure he is actually saving the gold. How many of those fine particles have slipped over the edge of the pan? For the new prospector, probably a high percentage.

Time is also to be considered. The hopper of a mini-rocker holds a pan or a shovel full of gravel. Even a new prospector can wash a mini-rocker load in less than two minutes. This is over 30 pans an hour. An experienced prospector can do much better. The new prospector with a pan generally finds it takes him several minutes to wash his sample. When he is done he is plagued with the thought he may have let some of the gold wash over the edge of the pan into the stream.

The mountain streams where gold is found are generally icy cold. With both hands in the water the fingers become numb after a time and painfully cold. The combination of the uncomfortable, crouched position with your knees on the rocks, plus the cold, aching hands is enough to discourage even the most enthusiastic prospector. With the mini-rocker there is no reason to put your hands in the water and you work in natural positions or even sit down if you wish.

Working Drawings of Mini-Rocker

Following are exploded views of the hopper and trough of the mini-rocker. These are separate drawings and show the individual parts, each of which is numbered and the dimensions shown. These numbers correspond to the parts list which is given in table form on a later page. You will note these parts are numbered 1 through 11 for the trough and 12 through 17 for the hopper. In addition, the dimensions are given on each part.

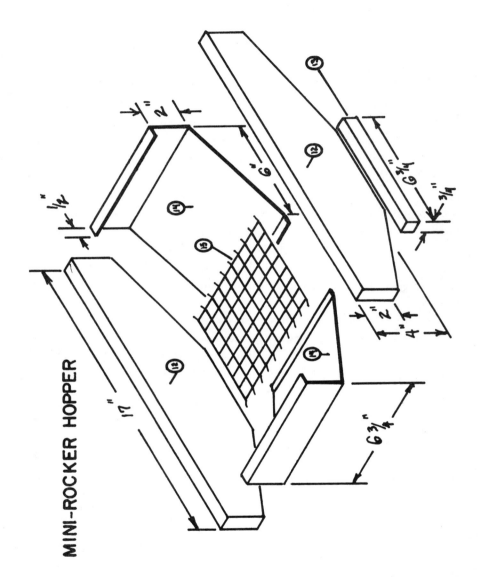

MINI-ROCKER HOPPER

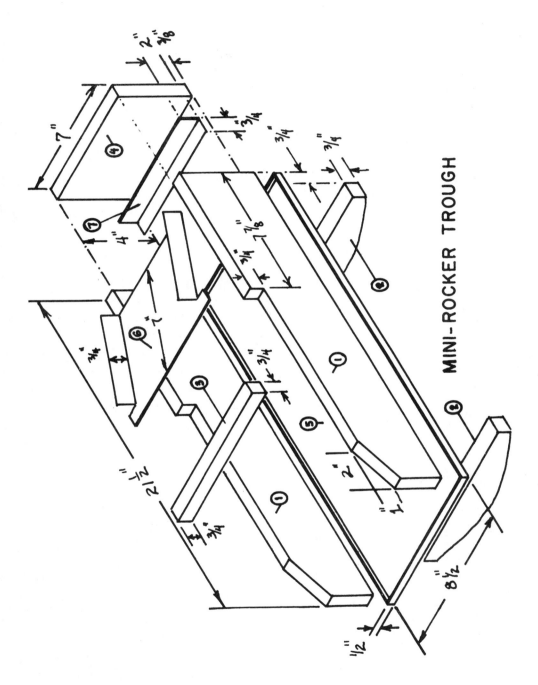

MINI-ROCKER TROUGH

How to Build Your Mini-Rocker

Like any construction job, the first thing you do after you have your plans and specifications is to get the materials for the job. I am going to assume you know how to use the basic tools of woodworking such as a saw, hammer, plane, square, screw driver, tape measure, and a bench to work with.

At this point I suggest you look ahead in this book to the page which gives the parts list for mini-rocker and instructions for assembly of parts. The dimensions of each piece are given along with explanations of what it is used for, and the kind of material required.

Wood

From this parts list you can probably figure out the amount and dimensions of the lumber you will need. You may have scrap around that will suffice. It doesn't have to be new material. Otherwise you will want to buy a piece of 1 × 6 long enough to make the parts. It is necessary to begin with 1 × 6 in order to secure the full four-inch material required. Regular 1 × 4-inch lumber is only 3½ inches wide and ¾-inch thick. When you add up the lengths of the several pieces of wood ¾ inch by 4 inches required you will find a ten-foot piece of 1 × 6 will be enough.

Metal

From the parts list you will note there are 3 pieces of metal in the trough and two pieces in the hopper, all #26 gauge, galvanized sheet metal. Since you probably don't have equipment for bending sheet metal, I suggest you purchase the pieces from a sheet metal shop, bent according to the view of the trough and hopper and dimensions in the parts list. You will note from the instructions for assembly the steps to follow in the assembly of the various parts. *These should be followed exactly as the sequence of the steps is very important.*

The metal bottom of the trough is 9½ × 22½ inches. This permits each edge to be turned up ½ inch at a 90 degree angle, leaving 8½ inches inside width and 21½ inches inside length. It is very important these inside dimensions are kept exact, as the side and end walls must fit into this metal bed (note steps 1 and 2). Be sure to use the 7-inch spacer block across the tail end of the bed. *This is a special piece not shown in the drawings.* You are to make it for this special purpose. It must be exactly 7 inches in length. When placing the assembled sides and rear end of the trough in the metal bed, push the front end firmly against the turned up front end of the metal bed first. Do this before you insert the 7-inch spacer. Then press the spacer in place toward the front end and when everything is firmly in place, nail the metal onto the wooden sides. You may then turn the trough upside down and bend the protruding portion of the metal bed up against the rear end piece and nail.

The baffle stock is 8½ inches wide which permits ¾ inch to be turned up on each side. To permit a ¾-inch lip of the metal to rest on the ¾-inch cross-piece, it will be necessary to make a cut in the metal to allow the 6¼-inch sides of the baffle to bend up. This is a 90 degree angle. (Note step #7 of assembly instruction and the exploded view of the trough.)

Parts List for Mini-Rocker

ITEM	QUANTITY	SIZE***	DESCRIPTION	
TROUGH		INCHES		
1	2	¾ × 4 × 21½	Side wall	wood
2	2	¾ × 2 × 8½	Rockers	wood
3	1	¾ × ¾ × 8½	Cross-piece	wood
4	1	¾ × 4 × 7	End wall	wood
5	1	9½ × 22½ × 26 gauge	Trough bottom	galv. steel
6	1	7 × 8½ × 26 gauge	Baffle	galv. steel
7	1	1½ × 7 × 26 gauge	Deflector	galv. steel*
8	1	7 × 19¾	Riffles	plastic mat**
9	20	2-d	Nails	galvanized
10	10	#9 × 1½	Wood screws	round head
11	1	¾ × 3 × 7	Spacer	wood
HOPPER				
12	2	¾ × 4 × 17	Side wall	wood
13	2	¾ × ¾ × 6¾	Legs	wood
14	2	6¾ × 9 × 26 gauge	Sloping bottom	galv. steel
15	1	6¾ × 6½ × ½" mesh	Screen bottom	hwd. cloth
16	20	2-d	Nails	galvanized
17	4	#9 × 1½	Wood screws	round head

*This part is to assure that no sand etc. gets under riffle mat.

**Plastic mat for riffles: The following described material forms very effective riffles and is recommended as being the best available. Green vinyl, runner material—manufactured for floor mats, welcome mats, aisle runners etc. Designated in dealer's catalogues as "Tinex Turf Pattern", Indoor-Outdoor Grass. Order number: F151-372
            L5003-1

Comes in 30-foot rolls, 27 inches wide, handled by most lumber yards, hardware stores, floor covering retail stores.

Have retail store dealer cut a 7-inch strip—exact measurements, off the end of his 27-inch wide roll. You will have 6 inches left.

(If impossible to get this exact pattern, try other mat material with ridges or corrugations to act as riffles. However, the above described is the most effective we have found for the purpose.)

Note: Plastic mat is not shown in the drawing of the hopper.

***All dimensions are in inches.

Instructions for Assembly of Parts of Mini-Rocker

STEPS	INSTRUCTIONS

Trough

1. Put metal bed of rocker down first on work bench.

2. Place the assembled side walls and end wall piece on top of metal base. Put 7-inch spacer block across tail end of bed to assure exact width before putting in screws and nails. (Six nails on each side, through the metal into edge of sides. One nail through metal into each end of side.)*

3. Attach rockers on each end of bed. (Two screws, countersunk and two nails in middle through metal bed into each rocker.)

4. Place deflector with ½-inch clearance above metal bed. (Three nails at each end and in middle.)

5. Be sure 7-inch spacer block is across tail end of bed to insure exact width before going to next step.*

6. Drill holes for screws into ¾ × ¾ inch wood piece and through baffle lip—slide into place and insert screws into top edge of sides.

7. Adjust baffle so lower edge clears metal bed 1⅜ inches—put nails in baffle flanges to secure baffle in position. (Two nails on each side.)

8. Bend 6¾ × 9 inch metal pieces to shape as shown in drawing.

9. Assemble two sides and two metal parts.
    Two nails through metal into end of each side.
    One nail through metal into top of each side.
    Three nails through metal into lower edge on each side, but leave out the lower four nails until last step (below)
    Slip screen between metal parts and side piece and put in last four nails.

10. Attach legs. Drill small holes for screws in legs, then attach to sides and insert screws. This secures the screen bottom of hopper.

---

*The reason for having the exact width of seven inches maintained is to assure that the 6¾-inch wide hopper can slide easily into the 7-inch trough for carrying and back packing. This is one of the unique features of the mini-rocker.

Now that you have your mini-rocker built and ready for use, the next thing is to get acquainted with it. It is a valuable tool that will go a long way in helping you become a successful gold prospector. But like any good tool it is necessary to learn how to use it properly. Fortunately, learning how to operate the mini-rocker is both simple and easy. In the instruction for using the mini-rocker a step-by-step method or system that Bill Hinsen has found to be the most satisfactory, is given.

Practice

There is no reason to wait until you are on a possible gold-bearing stream or placer to give your mini-rocker a try-out and teach yourself how to use it. Most any kind of river-washed gravel will do to use for practice. The gravel from any stream, excavation or stockpile will suffice providing it is river-washed (rounded pebbles, sand and gravel). Any natural source of river-washed gravel will contain sand along with the gravel and most likely some black sand. The black sands generally contain iron compounds which are several times heavier than other sand so will be found on your mat of the rocker after the lighter sands have been washed over the end of the bed. Gold is even heavier than the black sands so it is always found beneath the black sand. So find yourself a bucket of gravel and follow the instructions.

Of course you need more than gravel to carry on your practice testing. Put a half dozen whole BB's and several cut up ones in the gravel to simulate gold. Count the pieces and account for each piece on cleanup. You will also need water and a place to work.

With very little effort and additional equipment, you can set yourself up a suitable place to carry on your practice work. But rather than outlining a suggested set-up here, I will refer you to a later discussion on the use of the mini-rocker under desert conditions. You will find the method suggested there essentially the same as you will need for practicing.

Instructions for Using the Mini-Rocker

1. Set the rocker on a piece of plywood, a board, or a flat rock with the head end approximately an inch higher than the foot.

2. Place the hopper on the head end of the rocker, crosswise with the bed.

3. Fill hopper level full with gravel.

4. Make a 1-inch hole down through the gravel to the screen.

5. Fill the dipper with water and pour the entire dipper full into the hole at once, washing it out and making it bigger. Do not rock the cradle but let it set a couple seconds so the water can seep out into the gravel.

6. Fill dipper the second time and pour in all at one time. Start

rocker action very gently. Fill dipper the third time and pour into hopper and continue gentle rocking.

7. Fill dipper and pour into hopper, 4, 5, 6, 7 times and roll the rocker briskly while pouring.

8. Fill dipper several more times and pour into hopper and roll the rocker vigorously until the screen in the hopper is clear of all light sand, gravel, and clay, leaving only heavy, coarse gravel.

9. Check coarse gravel for possible large nuggets and dump the gravel out of the hopper.

10. Remove the hopper and slowly pour one full dipper of water onto the baffle plate while rolling the rocker gently to flush the riffles bed. The object is to wash away any light sand and gravel. There will be some black sand remaining in the last ⅔ of the riffle bed. The gold will be found on the foreward part of the plastic riffle bed under the metal baffle. The concentrates, consisting of the black sands and gold, should normally be as much as a tablespoon full to a half cup for one load of gravel.

11. This completes the separation process of one hopper full of gravel. It will take a few loads to get acquainted with your mini-rocker. It may take a few minutes for each of your first few loads until you are familiar with the several steps, but after your trial runs you should be able to process a hopper full in less than two minutes.

12. You do not have to clean up after each load. Up to a dozen loads can be processed before you clean up. After running 10 or 12 loads, the volume of concentrates will normally be approximately a cup full.

13. Repeat the steps suggested above until you have run 10 or 12 loads or your concentrates have built up to approximately a cup full. Get ready for the thrill that all prospectors look forward to—THE CLEANUP!

Suggestions on the Cleanup

Steps 12 and 13 above prepare you for the cleanup. This refers to the handling of the accumulation of black sands after a "run" with the rocker. You may want to clean up after running only one load of gravel through if you are load testing or, as suggested in steps 12 and 13, you may run as many as 12 loads through before you make your cleanup.

The Cleanup

Following are step-by-step directions of how to proceed with the clean-up of the concentrates from the mini-rocker.

1. Remove hopper.

2. Raise head end of rocker by placing a ¾-inch board under the front rocker.

3. Fill dipper full and pour carefully onto metal baffle, and at the same time roll the rocker gently. One gallon through is enough. Plan to work it down to approximately a cup full of concentrates.

4. Pull plastic mat out of rocker bed by lifting the tail end of the mat and pull out carefully. Wash the mat out in a gold pan ¾ full of water.

5. Rinse out all sand and gravel from rocker box, especially under baffle, into the gold pan so the plastic can be replaced.

6. With the concentrates now in the gold pan, while working over another pan for safety, you may be able to wash a little more of the lighter sands out. If not, place the pan with the concentrates in a warm place to dry out. When dry, you can easily pick out the larger pieces of gold with your tweezers and put them in a plastic flask or bottle. The fine gold and black sand concentrates are placed in a separate metal or plastic container for further attention.

At the cleanup the plastic mat is carefully removed and the concentrates washed off by dipping it upside down in water, in a gold pan.

Rinse out all sand and gravel from rocker bed into the gold pan, especially under baffle so the plastic mat can be replaced.

Black Sand

A beginning prospector may be disappointed in his first cleanup when he finds he has what appears to be only a small quantity—perhaps a cup full of black sand. These are his concentrates. But he need not be disappointed—because—even if the concentrates are rich with gold, it will still appear to be just a pile of black sand. The gold is so much heavier than the black sand that it is always on the bottom, covered up with the black sand so you can't see it. Not until you have your concentrates in a gold pan, perhaps a quarter full of water, and work it around with a certain circular and rhythmic motion that washes the black sand on ahead of the gold, will you see the long string of gold dust, flakes and tiny nuggets. How to separate the gold from the black sands is discussed in chapter IV, "Recovery of Placer Gold."

Desert Prospecting With the Mini-Rocker

It is well known the water supply in the desert is a serious problem. The lack of water to test the samples with a pan in a running stream seemed almost to prohibit prospecting under desert conditions. A mini-rocker can help solve this problem.

A portable, improvised washing plant can be put together. It consists of; a wash tub, a deck, the mini-rocker, two sawhorses, barrel or five gallon cans, and water.

The deck consists of a piece of ¾-inch plywood, 3 feet long, 2 feet wide on one end and 16 inches on the other, with 1 × 4 inch sides and the wide end the same. The deck is set up on the sawhorses so that the narrow end is over the edge of the tub sitting on the ground. Give the deck a slope of approximately one inch. A ¾-inch strip of wood is fastened across the lower end of the deck for the front rocker to ride against to keep the whole machine from creeping forward. Place the strip so the end of the bed sticks over the edge of the tub.

The water supply is brought to the site in a barrel or cans and used over and over to wash the samples. The sawhorse legs (25 inches long) will support the deck at a convenient height for a man sitting on a 12-inch stool. From this position he can reach into the tub with the long-handled dipper with one hand and rock the rocker with the other. Be sure the barrel and cans are free from any gasoline or oil as any contamination of this kind will cause the fine gold to float out. Add detergent to prevent this.

Desert prospecting with the mini-rocker.

Moss Gold

In some sections of the country moss grows in the streams. It hangs in long strings on the sides and bottom. The fine mesh of tendrils catches and holds the fine gold as it washes down. Some prospectors are aware of this and try to wash out and recover the gold by various means. It has always been a problem to find a way to do this. The mini-rocker has proven to be very helpful with this problem. There are three different ways to it. The first way is to place a small batch of moss in the hopper and pour water over it as you rock the rocker vigorously. Mix and stir the moss with your hands at the same time. The slit and gold will wash down through the screen on the baffle and mat. The second way is to simply wash the moss out in a bucket or tub by lifting it up and down and shaking it in the water. The silt, gold and black sand will fall off and settle to the bottom. The water is poured off and the sediment gathered up and put through the rocker. The third way is to remove the hopper and tilt the tail end of the rocker bed up and fill it with water. Place the moss in the water and work it around with a scrubbing motion on the mat to break up and separate the moss fibers, thus releasing the gold and silt which will fall to the mat.

In all three methods, the cleaned moss is discarded and the sediment is washed off the mat into the gold pan. The sediment can be stirred with enough water to form a thin slurry which is poured onto the baffle while the rocker is being rocked. Follow the same procedure as outlined in handling the concentrate and the cleanup.

Garnets and Sapphires

Garnets and sapphires are sometimes found in the same gravel with gold. They are not as heavy as gold but are heavier than most of the other gravel in the aggregate. At some point, for instance, at step #7 of the instructions for using the mini-rocker, observe and check the riffle bed to see if there might be either sapphire or garnet specimens. They can be easily identified by their distinctive colors, the garnet being a deep red with a five-sided crystal formation. The sapphire has its bright blue, yellow or other colors. Pick these out if present and put in a plastic bottle for further identification and disposition.

Your mini-rocker is a simple but highly efficient tool. It is fast, accurate and positive in accomplishing the purpose for which it was designed—to test your samples and to produce gold. As you become experienced in its use you will come to realize its merit more and more.

It is the lineal descendent of the old log washer, an idea that was conceived by one of the early prospectors in the south during America's first gold rush in the early 1800 era. This was many years before the famed California gold rush. Those southern miners who later joined the rush to California brought the idea with them. It was quickly accepted and improved, and the historic California washer was the outgrowth of the original log washer. Along with the sluice box, the rocker or California washer became standard equipment of the prospectors and placer miners of those early days. It was efficient and practical as a means of producing gold and remains so today in the gold rush of the 1980's.

It remained for someone to adapt this efficient production tool to the problem of testing a sample—to detect the presence of gold in river-washed gravel. This had always been the exclusive function of the gold pan. The mini-rocker, developed by Bill Hinsen, has accomplished this purpose and combined it with the means to also produce the gold when found.

If you have built your own mini-rocker according to the information given, you now have a tool that will both detect and produce gold for you if given the opportunity. It is up to you to put it to work!

Chapter IV
Recovery of Placer Gold

Panning has a romantic and rustic appeal, but there are other, more productive, simple methods of gold recovery. We'll discuss a couple of these.

The Cradle or Rocker

The cradle or rocker was widely used by the early miners and it is still a valuable tool for recovering placer gold from gravel. The cradle is light-weight, easily and cheaply constructed, durable, readily transported, and can be operated by only one or two people. It will handle three to five cubic yards of gravel in a ten hour day and it will save all but the very smallest of gold particles. Details of its construction are found on page 47. Its operation is simple. A stream of water equal to the flow from a garden hose is fed into a screen-bottomed hopper containing a load of gravel. At the same time, the device is vigorously rocked back and forth. This action washes and sorts the gravel and sand containing the gold. The larger gravel is thrown out of the hopper by hand to allow for the addition of new material. The sand and gold drop through the screened bottom of the hopper into a riffled trough, the slope of which is regulated by raising or lowering either end of the bed plate. At the end of a day's work, any gold that has accumulated in the riffles is obtained by removing the riffles and flushing the sand down the trough into a tub. The gold is recovered by panning this sand.

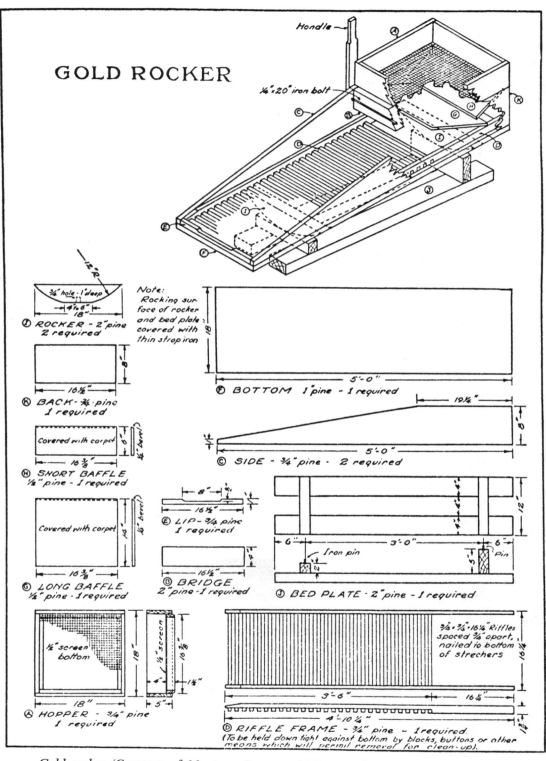

GOLD ROCKER

Handle

¼"·20" iron bolt

Handle

Note:
Rocking surface of rocker and bed plate covered with thin strap iron

½"·2"

¾" hole · 1" deep
4½·6"
18"

(I) ROCKER - 2" pine
2 required

8"
16½"

(K) BACK - ¾" pine
1 required

Covered with carpet
16⅜"
6"
½" bevel

(H) SHORT BAFFLE
½" pine - 1 required

Covered with carpet
16⅜"
14"
½" bevel

(G) LONG BAFFLE
½" pine - 1 required

8"
16¼"

(E) LIP - ¾" pine
1 required

16½"

(B) BRIDGE
2" pine - 1 required

18"

(F) BOTTOM 1" pine - 1 required
5'-0"

19¼"
6"
5'-0"
½"

(C) SIDE - ¾" pine - 2 required

12"
6"
3'-0"
6"
4"·4"·4"
Iron pin
Pin
5"

(J) BED PLATE · 2" pine - 1 required

½" screen bottom
18"
18"
½" screen
16¾"
1¼"
5"

(K) HOPPER - ¾" pine
1 required

⅜"·¾"·16¼" Riffles
spaced ¾" apart,
nailed to bottom
of strechers

16¼"
3'-6"
16¼"
3'-6"
4'-10¼"

(D) RIFFLE FRAME - ¾" pine - 1 required.
(To be held down tight against bottom by blocks, buttons or other means which will permit removal for clean-up.)

Gold rocker (Courtesy of Montana Bureau of Mines and Geology, Memoir)

It is important that the water and sand be forced to flow *over* the riffles. If the mixture flows under the riffles, it will carry the gold out the end of the trough. A piece of canvas, blanket, or other fabric can be put between the riffle frame and the bottom of the trough to prevent this, with the riffle frame wedged tightly against the bottom. The baffles are covered with carpet, burlap, or some similar material. They serve to spread and feed the gravel evenly down to the riffles.

The Sluice Box

The sluice box is an even easier piece of equipment to make than the cradle and it is probably more widely used. (See the construction details illustrated on page 50.)

Ordinarily, each sluice box section is 12 feet long and 12 inches wide on the inside, with two sides and the bottom forming a trough. The sluice box is made of one or two inch material, depending on whether it is to be used as a portable unit or as a permanent one. The riffles are most important. They are held in place with a frame that holds them tightly against the bottom so that no water can pass under to carry the values out of the box. Square pieces of wood placed crossways in a frame are the most commonly used riffles, but other types frequently used include poles placed crossways or lengthways in the box. A layer of larger rock can even be used.

To keep the box from sagging, warping or otherwise losing its shape—which results in leaks and inefficient operation—the box must be properly supported and braced. (See Figure 5.) The slope of the box is very important and varies somewhat depending on the amount of water flowing into the box and the nature of the aggregate being washed. Six to twelve inches is an average gradient for a 12 foot box. First level the box with a carpenter's level; then raise or lower one end to get just the right amount of slope for your needs.

Although both the cradle and the sluice box are simple devices there are a number of details to consider, not only in their construction and operation, but also in the type and arrangement of the riffles. If you intend to build and operate either a cradle or a sluice box, it is recommended that you consult one or more of the references listed in the back of this book. One special source is Miscellaneous Contribution No. 13, published by the Montana Bureau of Mines and Geology in Butte, Montana: *Practical Guide For Prospectors and Small Mine Operators in Montana*, by Koehler S. Stout.

Both the sluice box and the cradle require a water flow, so it would be wise to confine your search for placer gold to creeks and rivers or to areas where water can easily be diverted or pumped to your washer. Old flumes and ditches along the mountain-sides of the West remind us of some miner's work to bring water to his diggings in an out-of-the-way dry spot.

Mechanical Equipment

One device that the miners of the 1930's found to be of great value that the prospectors of an earlier day would have given their eye teeth to have owned is a small, portable, gas-engine-powered pump with a long flexible hose similar to a fire hose. With this, they could set up their sluice box near a rich bench or bar some distance from the creek or river and pump the water to their box. It saved them a great deal of work in transporting the pay gravel to a sluice box near the creek.

In this modern age of power equipment the placer miner can be in business on practically any scale, ranging from a small, portable pump, to a much larger operation using a bull-dozer, power shovels, and draglines to remove the overburden (the useless material above the pay gravel) and deliver the gold-containing gravel to a trommel or a large, mechanical washing plant. Of course, this type of operation requires a large amount of money; still, several hundred cubic yards of gravel can be washed in a day and a much lower grade of pay gravel can be handled than in a hand operation.

The Floating Dredge

The huge, unsightly piles of gravel waste in the beds of many creeks in the western states are evidence of the use of enormous floating dredges to work the gold-bearing gravel back in the early days. These monsters inched their way up the creek bed, gobbling up the earth and discharging the gravel behind them. In their day, they were fairly efficient machines. One by one, however, they fell into disuse due to increased operating costs and the fixed price of gold. Some were

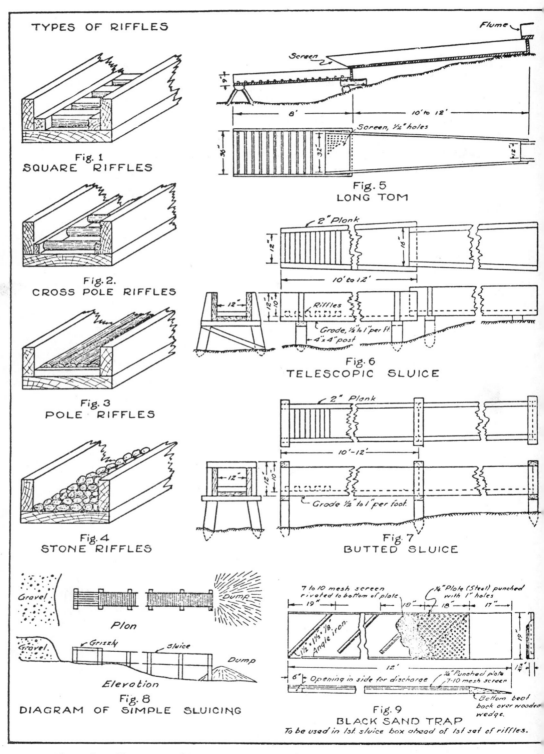

TYPES OF RIFFLES

Fig. 1
SQUARE RIFFLES

Fig. 2.
CROSS POLE RIFFLES

Fig. 3
POLE RIFFLES

Fig. 4
STONE RIFFLES

Fig. 5
LONG TOM

Fig. 6
TELESCOPIC SLUICE

Fig. 7
BUTTED SLUICE

Fig. 8
DIAGRAM OF SIMPLE SLUICING

Fig. 9
BLACK SAND TRAP
To be used in 1st sluice box ahead of 1st set of riffles.

Sluice box (Courtesy of Montana Bureau of Mines and Geology, Memoir No. 5).

Floating Yuba dredge. This dredge, located on Prickly Pear Creek near Jefferson City, Montana, between Helena and Boulder on U.S. Highway 91, was built 4¼ miles from its present site in 1938. It operated successfully until 1958, when it was closed down due to the reduction of gold content of the gravel. During its period of operation it produced more than 65,000 ounces of gold—worth two and a quarter million dollars at $35 per ounce.

The maximum capacity of the dredge is 8,000 cubic yards of gravel per 24 hour period. It is powered by electricity and with minor repairs it would be in operable condition. With the increased price and interest in gold it may again serve a useful purpose, probably in another country where ecological standards are not as high as in the United States.

dismantled and shipped to other countries where operating costs were lower, but others have remained on the site of their last operation as ugly reminders of that period in the history of placer gold mining.

Typical of these huge floating dredges is one near the highway in a pond on a small creek near Jefferson City, Montana, between Helena and Boulder. A picture of this representative of these old dredges is found on page 51.

The Modern Portable Floating Dredge

There is a modern, lightweight, portable version of the floating dredge which is both inexpensive and efficient. It combines scuba diving gear; a rubber raft or other platform; a light, efficient gasoline engine; suction equipment; and a portable sluice box. These dredges work on the same principle as a vacuum cleaner and can be used in ponds, rivers, or creeks.

Initial underwater exploration can be done with a face mask and snorkel, to avoid setting up the floating dredge if the location proves to be unfavorable. The objective of the floating dredge operator is the same as the crevice man, except that he is working under-

Modern floating dredge. Small floating dredge on Sucker Creek, which is a branch of the Illinois River near Caves Junction, Josephine County, Oregon, in August 1974.

water. In addition to suction equipment, he often uses crevice tools
to work the deep and narrow cracks where the gold is most likely to
be found. The vacuum hose of the suction dredge picks up and delivers
the sand and gravel from the crevices and other traps to the upper end
of the portable sluice box on the floating platform.

This method of mining placer gold was not even dreamed of by the
old prospectors of the early days or miners in the Depression days of
the 1930's. With equipment such as this, the modern gold seeker has
an enormous advantage over the miners who have gone before him.

For anyone interested in more information on the floating dredge,
one of the most complete sources is the June 1972 issue of *California
Geology*, available from the California Division of Mines and Geology,
P. O. Box 2980, Sacramento, California 95812.

TYPICAL GOLD - DIVING OPERATION

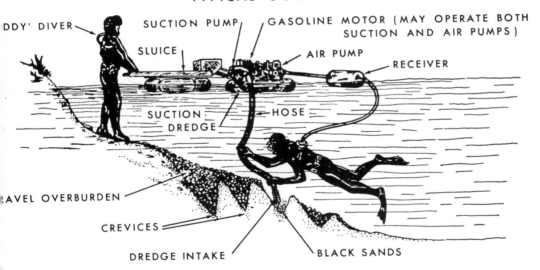

*In the typical gold diving operation, the gasoline-engine powered air
compressor and suction pump can be located on shore or mounted on flooats so
they can be towed by the diver as he works under water. (From "Diving for
Gold" by William B. Clark, California Geology, April, 1972, California
Division of Mines and Geology.)*

The Floating Jet Dredge

The modern suction or jet dredge uses the Venturi tube principle. A motor driven pump on the surface feeds water under pressure to the head of the Venturi tube where suction is created by the stream of water passing swiftly into the larger tube. The created suction, or vacuum, in a flexible tube is directed by the operator to crevices and cracks and along the face of the bedrock of the stream, or pond. The intake from the suction is run into a sluice box where the riffles collect the gold as the gravel and water passes through to the discharge end of the sluice box.

This pipe-like device ranges from 4 to 8 feet in length and weighs up to 20 pounds. It is usually made of galvanized sheet metal and the curved or intake end is of stainless steel or other resistant material.

The jet dredge and the suction pump dredge both use "suction," or a partial vacuum principle, to suck up the gold bearing sand and smaller gravel from the crevices and other likely places on the streambed where gold may have accumulated.

Unlike the jet dredge, the suction pump dredge creates the partial vacuum by the use of a vacuum air pump driven by a small, portable gasoline engine which may also operate the air compression pump.

Deep Channel or Buried Placers

Another type of placer in the gold country is an ancient streambed sometimes called a deep channel, or buried placer. If a creek or river changed its course due to a rock slide or some other natural event, the dry streambed containing gold-bearing gravel soon filled with weathered rock and debris. Even major earth changes may have occurred; some buried placers are now covered by clay and lava. Some of the most productive placer mines in the Mother Lode section of California were buried placers. These deep channel or buried streambed placers were recognized to be one of the most productive sources of placer gold when the mines were closed down during World War I. Many potentially valuable deep channel placers are still lying undiscovered, waiting for some knowledgeable and industrious prospectors to find them.

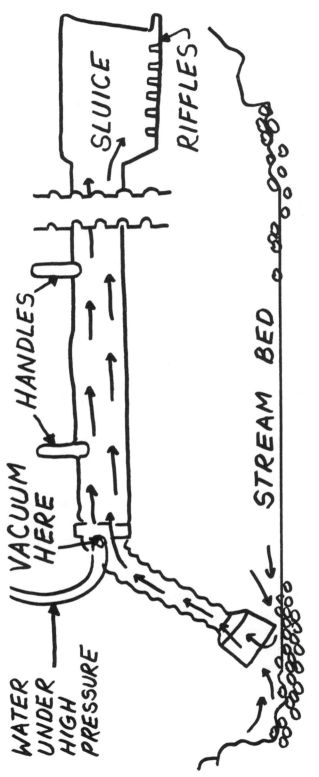

JET DREDGE

SLUICE

RIFFLES

HANDLES

VACUUM HERE

WATER UNDER HIGH PRESSURE

STREAM BED

Jet dredge.

Working the Deep Channel Placer

To work a buried placer, it is often necessary to tunnel into the gold-bearing gravel along the slope of a hillside. This is termed drifting or drift mining and is common in hard rock mining. This tunnel or adit is used to haul out the gold-bearing gravel and to provide drainage as well, since water is often present in a drift. Sometimes the pay gravel is hoisted up through a shaft to the surface, and the adit is used only for drainage. A shaft inclined at 45 degrees is often used by present day operators as the gravel can be moved up the inclined grade with less expensive equipment and with greater safety. In the event of equipment or power failure, the miners can walk up the inclined shaft to the surface.

A source of water is necessary to wash the sand, gravel, and clay away from the gold. Enough seepage or spring water is often available in the mine for washing the gravel, if a system is provided to circulate and re-use the water, such as a series of settling basins or ponds. This system is practical and it meets ecological and debris law requirements.

Machines Work for the Modern Miner

Horse, mule, burro, occasionally oxen, and human muscle were the principal sources of power for the miner of a hundred years ago. Even the Depression miners of the 1930's were limited in their choice of machines to assist in their search for, and recovery of, gold. But now, 45 years later, many machines developed and built for road building, dirt moving, hauling, or other types of mining are being altered and adapted to the various types of gold mining.

Following are pictures taken of a trommel gravel washing plant which operated near Nevada City, Madison County, Montana, during the summer of 1974. This is not necessarily a typical small miner operation but shows the trommel being used as a washing plant for placer gold. In this case the water supply was limited to the small, mid-summer flow from springs feeding into Brown's Gulch. This water was caught in a pond below and pumped back to the trommel. The

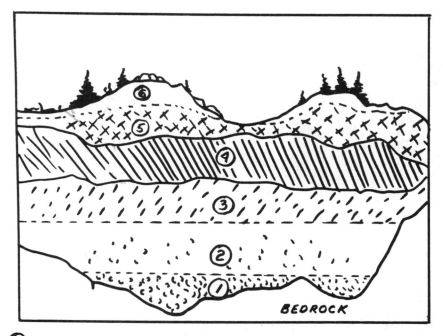

① PAY GRAVEL. 1 OZ. PER YD.   LOWER SECTION
② PAY GRAVEL, ½ OZ. PER YD.   UPPER SECTION
③ PIPE CLAY
④ LAVA
⑤ SUB SOIL
⑥ TOP SOIL

Deep channel placer—cross-section.

GMG *Assay Office*

A Division of GOMIL CHEMICAL CO.

MINERS' EXCHANGE BUILDING

432 WEST MAIN STREET - QUINCY, CALIFORNIA 95971

PHONE: 916-283-2280
CABLE ADDRESS:
"TRANSPHERE"
QUINCY, U.S.A.

MEMORANDUM OF ASSAY

MADE FOR George A. Nugent .... DATE .... Dec. 21, ..., 196 74

SAMPLE NO.	PER TON OF 2000 POUNDS AVOIRDUPOIS								COPPER, OR			LEAD, OR			TOTAL	
	GOLD				SILVER											
	AT		PER OUNCE		AT		PER OUNCE		AT		PER LB.	AT		PER LB.		
	OZS.	100'S	$	CTS.	OZS.	100'S	$	CTS.	%	$	CTS.	%	$	CTS.	$	CTS.
1. Outcroping	0	02	189	25											3	77
2. Wade ?	C	01	198	25											1	88
3. Tailings	C	04	189	25											7	53

ASSAY NO. ....

CHARGES $ 7.50 Paid WEM

BY ....
WILLIAM E. MILLER, ASSAYER.

CHEMISTRY *Touches* **EVERYTHING**

Assay report. The No. 1 sample was taken from an outcrop of a deep channel placer in Plumas County, California. It was on the top six inches of a 72 foot section of fine white quartz and was taken to identify the outcrop rather than to find values. It was a surprise to find $3.77 worth of gold in a sample that close to the top of the section.

The Wade Sample was from the upper section of a deep channel placer from a different mine several miles distant.

The Tailings (or No. 3) sample was from the waste pile of the original sluice box workings of 1923. They have been reworked twice since they were originally mined and washed. This sample is also from a Plumas County, California, deep channel placer.

amount of water available is very important in determining the type of washing plant to be used. A system of ponds that permits sediment to settle precludes pollution of the stream and makes it possible to operate a washing plant on a small supply of water.

New and more efficient gravel washing plants are being designed and built to meet the greatly increased demand for equipment of this kind—stimulated by the five-hundred percent increase in the price of gold and the passage of the private gold ownership Act of August 1974.

Centrifugal Gold Concentrator

Anyone who has had experience on the farm turning the handle of an old-fashioned cream separator knows that the lighter cream is separated from the heavier skim milk by the use of centrifugal force. The milk is released from the tank above into the center of a revolving bowl containing a number of disks. If the bowl is turning at the proper speed, the comparatively heavy skim milk is thrown to the outside of the disks, collected, and flows through a spout into a bucket. The cream, being noticeably lighter, is forced to the center of the disk and bowl, and into a spout leading to another receptacle, thus completing the separation process.

In the centrifugal gold concentrator the principle of centrifugal force is used in a way similar to the cream separator. The heavier gold is separated from the sand and fine gravel as it is fed into the center of the revolving machine with a continuous stream of water. The inside of the bowl has a removeable rubber liner, contoured to form riffles, which catches and holds the gold after it is thrown to the outside of the bowl by centrifugal force when the bowl is revolving at the proper speed. The waste water, sand, and light gravel boil up over the sides of the revolving bowl and are caught by a metal shield which surrounds the bowl and then flows down into a trough leading to the discharge pipe, thus completing the separation process. If the operator wishes to

Rich, gold bearing gravel was mined from this gulch near Nevada City, Madison County, Montana, in the summer of 1974.

Brown's Gulch is part of the Alder Gulch—Virginia City, Montana area. "ALDER GULCH (includes Junction, Nevada, Fairweather, Highland, Pine Grove, Summit, and Brown's Gulch) Virginia City, 8 miles east of Alder, N.P.R.R. Discovered in 1863, has yielded over $50,000,000 by sluicing, drifting, and dredging, in Alder Gulch and other tributaries. Active in 1932, 1933, and 1934. Possibility of buried channels under basalt flows should be investigated." From Reprint of Part II of Memoir 5, Placer-Mining Possibilities in Montana, by O. S. Dingman, (1971) Bureau of Mines and Geology, Butte, Montana.

A portable trommel used as a gravel washing and gold recovery plant on Brown's Gulch near Nevada City, Madison County, Montana, July, 1974.

The cleanup at the trommel is always a job. More than a pound of coarse gold was recovered on this one on July 1, 1974 on the gulch near Nevada City, Madison County, Montana.

The cleanup (gold bearing sand) from the trommel is flushed into the scoop of the front end loader and then shoveled into a centrifugal gold concentrator where much more of the sand is washed away. The concentrate can then be panned by hand to separate the last of the sand from the gold.

Side view of centrifugal gold concentrator. The stationary shell, supported by four legs, protects the revolving bowl inside which is driven by an electric motor or gas engine.

Top view of centrifugal gold concentrator. The inside of the bowl has a removable rubber liner contoured to form specially designed riffles which catch and hold the gold after it is thrown to the outside of the bowl by centrifugal force. (Photographs courtesy of Duke's Manufacturing Co., 315 Colorado River Boulevard, Reno, Nevada 89502)

check the efficiency of the bowl he can run the discharged material into a sluice box.

The concentrator has a capacity of 4 to 5 cubic yards per hour. It can be used with material from a placer, tailings, or dump or to further concentrate the gold in the cleanup from a trommel or other type of washing plant.

The Concentrates

It may be a surprise to the beginning prospector to find that regardless of the method or equipment used to produce the gold from the sand and gravel—whether it is a pan, sluice box, rocker, trommel or a washing plant—the end product is not a quantity of gold nuggets, coarse gold, and gold dust. Instead it is a pile of black sand! Because of the extreme weight of gold it lies hidden beneath the black sands. Together, they are known to the placer miner as concentrates.

The reason the black sands are found so consistently with gold is they are both heavier than the other sand and gravel in which gold is found. The black sands consist of different minerals depending upon their origin, but are approximately twice as heavy as the other sands. For instance, magnetite has a specific gravity of 5.19; hematite, 5.26; carundum, 4.02; galena, 7.4; compared with 2.65 for the ordinary silica-type rock such as quartz. Platinum grains are frequently found, which are even heavier than gold and some of the black sands are actually gold that is stained or coated with some black substance that masks its identity. This accounts for the black sands presence, but the problem is to separate the gold from the black sands.

Specific Gravity

Before going into our discussion of black sands, gold and concentrates, let us define the term specific gravity, which simply means the weight of a substance compared with the weight of water. In technical terms the weight of water is one (1). One cubic centimeter of water weighs one gram. That is our basis for comparison. For instance, if we find a certain rock has a specific gravity of 2.5, this means it is two and a half times heavier than the same quantity of water. Gold has a specific gravity of 19.7 so we know it is 19.7 or nineteen and seven tenths heavier than water. With this information in mind, it is easier to understand the relationship between the black sands, ordinary sand, gravel, gold and water, so far as weight is concerned.

Since the problem is one of separating the black sands from the concentrate, let us proceed in a methodical, step-by-step manner.

First Step—Mechanical Separation

Assume you have already picked out the nuggets and coarse gold and have only the black sand and fine gold remaining. The following method relates to a small or sample size quantity. Some variation is used in larger, commercial scale quantities. Here is one way to proceed:

1. Spread it all out in your pan, a pie tin, or a plate, and set it in the sun or on a source of heat such as a radiator or stove to dry.
2. When dry, you can put a small quantity at a time on a piece of stiff cardboard and blow across the material, perhaps tapping the cardboard a little and you will find that a good share of the lighter weight sand will blow off at once. As you stir it around and blow again you will get rid of a little more but there will still be a quantity left. That which is remaining is probably one or more of the iron-containing black sands.
3. Since most of these compounds are magnetic they can be readily picked up with a magnet. Don't make the mistake of putting your bare magnet on the black sand. Some of the black sand will stick to the magnet in spite of everything you can do. Fortunately, this can be prevented by wrapping the magnet with a sheet of plastic before using. The black sand will cling to the plastic but will drop off as soon as the magnet is removed.
4. If there is still black sand of a non-magnetic type mixed with the gold, then your next step is to separate the gold with the aid of mercury by a simple process called amalgamation.

Second Step—Cleaning the Concentrate (Gold)

Before proceeding to the separation of the gold from the black sand, the concentrate must be thoroughly cleaned. The necessity for this preliminary step before attempting amalgamation cannot be overemphasized. In discussing the amalgamating powers of mercury in a placer gold washing plant with a consulting engineer recently, I was surprised when he said, "From my experience, the amount of fine gold crossing a mercury-coated plate that is actually recovered is comparatively small, perhaps 10 to 15 percent." He explained that the reason the percentage is so low is that practically all placer gold has a protective coating of some kind; sometimes pine oil, a manganese or iron oxide, or other substance. Unfortunately gold is often coated with these various substances including iron sulfide, magnesium or other inorganic substances which stain and discolor it. Any kind of coating will prevent amalgamation." Explaining further he said, "The answer is to treat the sand containing the gold with a cleaning substance before it comes in contact with the mercury."

This is extremely valuable information and has to do with the overall recovery rate from any placer gold recovery system or operation.

There are other ways of cleaning gold, but one very simple method employs the use of nitric acid. First, place the concentrated gold-bearing sands into a plastic gold pan or other container that will not be affected by strong acids. Cover the concentrates with about ½ inch of water and add the acid in small amounts until a slight boiling action occurs. This boiling begins when about a 90 percent concentration is reached. In some cases no dilution is necessary. Add straight acid. Mix the acid solution around for several minutes, making sure all the material is well saturated. Even better, let it stand a while. The acid solution is then poured off and the concentrates rinsed by dipping the pan in the stream. The gold is now ready for amalgamation.

Third Step—Separation by Amalgamation

Amalgamation is one of the oldest and simplest methods for recovering free gold from the black sand concentrates from placer mining. When clean particles of gold are brought into contact with mercury they unite, or merge into a single, new substance known as amalgam.

Amalgamation takes place when the two metals are brought into contact with each other by mixing or blending, and even the smallest particles of flour gold are gathered up and absorbed by the mercury. This special affinity for gold makes mercury an invaluable aid in separating the gold from the black sands. It takes only a few drops of mercury to gather the gold in the black sand found in an average pan of sample. The mercury should be agitated with the concentrate under water until every portion has been contacted and all the small globules and BBs of mercury are absorbed by the one ball of amalgam. It now contains all the fine and flour gold that would otherwise have been lost.

Fourth Step—Separating the Gold From the Amalgam

Phase 1. Quite frequently more mercury will be applied than is required for the amount of gold to be absorbed, in which case the ball of amalgam will actually be a mixture of free mercury and the amalgamated substance. In fact, it is desirable that an excess of mercury be used to assure that all the free gold in the sample is absorbed. However, in the interest of economy and the conservation of mercury, a preliminary step is required befor beginning the separation of the gold from the amalgam. This consists of removing as much of the free mercury from the amalgam as possible. This can be accomplished in several ways. Historically, the mercury-amalgam mixture was squeezed through a piece of chamois skin, or a silk cloth, but there are problems with this method. More recently, a syringe with a thin cotton pad in front of the plunger head is being used with success. The mercury-amalgam mixture is placed in the syringe and upon pressure, the mercury is squeezed through the cotton pad into a container, leaving the small, round, cake or wafer of amalgam in the syringe. The cake of amalgam is now ready for further treatment.

Phase 2. The next step is to be conducted outdoors as toxic fumes are produced. It is suggested you set up a stool, chair or small table in your back yard or other open area, then place an electric hot plate with a long extension cord attached to it, on the table. Place a pyrex dish ⅔ full of 25-65 percent nitric acid on the hot plate. Ordinary glass will break upon heating so the dish must be pyrex which is made to withstand high heat.

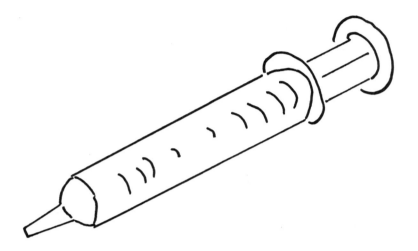

A Cattle Syringe, 25 to 65 cc size has proven to be a useful tool for the place miner. The mercury-amalgam mixture is placed in the syringe and upon pressure the mercury is squeezed through the cotton pad into a container leaving the small, round cake, or wafer of amalgam in the syringe.

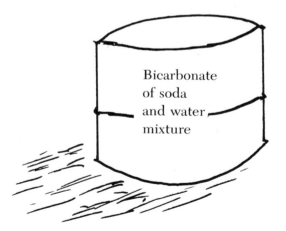

Bicarbonate
of soda
and water
mixture

You should have a half gallon or more of water and bi-carbonate of soda handy in case you get acid on your hands or clothes. If you should get acid on your hands, dip them in the soda water immediately.

You are now ready to place the cake of amalgam in the solution. Almost at once you will note a reaction taking place as fumes begin to rise. These fumes may not be fatal but are very toxic and if inhaled may make you sick and cause damage to your lungs, so stay away from them. Next, you turn the current on to the hot plate and let the solution boil for 15 minutes. What is happening is, the mercury from the amalgam is being absorbed, or taken into solution by the nitric acid and leaving the free gold in a natural form. At the end of 15 minutes turn off the current. After the dish has cooled and no more fumes are coming off, walk over and inspect your cake of gold. Pick it out of the solution with a pair of pliers and drop it into a cup of water with ordinary bicarbonate of soda in it to neutralize the acid on it. You will note the little wafer or cake of gold is not solid, but rather looks like a sponge.

After the mercury was absorbed by the nitric acid, it left empty spaces where the mercury had been, giving the remaining gold a sponge-like appearance—drab and rusty in color. In fact, it is called sponge gold and is ready to be melted down into a bar, or used for any purpose desired. Heating the sponge to blowtorch temperature will bring out the natural, bright gold color.

Incidentally, you should have a half gallon, or more, of water and a bicarbonate of soda mixture available. In case you get some of the acid on your hands, dip them in the soda water immediately.

Fifth Step—Recovering the Mercury

The remaining nitric acid solution now contains the small amount of mercury which came from the amalgam. This mercury can be recovered by placing a small sheet or bar of copper in the solution. The mercury will collect on the copper and can be scraped off and saved for future use.

The used nitric acid solution should be diluted and neutralized by adding soda water and flushed away. It must not be put through the plumbing system unless diluted and neutralized. The acid solution can be saved and reused again but nitric acid is so inexpensive that this seems unnecessary.

Objective Accomplished

What these five simple steps accomplish is: mechanical separation, cleansing of the concentrate, separation by amalgamation, separation of gold from amalgam, and recovery of the mercury.

In processing larger quantities of black sands these steps are accomplished by other methods appropriate to the quantities involved.

Put a pyrex dish ⅔ full of 25 to 65 percent nitric acid on a table in your back yard, or other open area. The fumes are very toxic. Take no chances.

Chapter V
Hard Rock or Vein Gold

Surely there is a vein for the silver and a place for
gold where they find it.

Job 28: 1

Prospecting is a tiring job that requires at least an elementary
knowledge of geology and considerable patience. The old timers may
not have known that their practical knowledge of where and how to
find gold was called geology, but they knew it produced results. They
had their fair share of patience—but even so, many prospectors have
walked away from their diggings, discouraged, only to have someone
else come along, dig a little deeper, and strike it rich.

Binoculars—A Prospector's Tool

In looking for faulted and mineralized areas in mountain country,
some of the old prospectors found a good pair of binoculars to be a
time- and labor-saving piece of equipment. They could then study
exposed rock over a wide area. Faulted areas, bands of colors, and
areas of broken white rock indicating quartz could be detected in this
way. Promising areas would then be checked out on foot, and the
quartz would be crushed and panned for gold.

Faults

Faults are breaks or fractures in rock which generally extend deep

70

Binoculars—a prospector's tool.

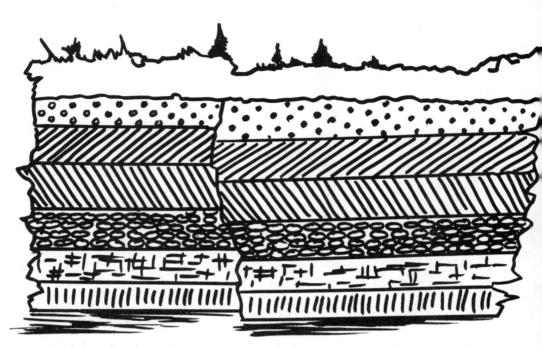

Faults are breaks or fractures in rock which generally extend deep into the earth.

into the earth. One side of the break will probably be lower than the other. This is called displacement and it is measured in feet and inches. The cracks and fissures resulting from the break provide a channel that permits molten minerals mixed with gas, water, and steam to rise to the surface. This material is under tremendous pressure and the cooling is often slow so crystals form. If the molten material contains quartz it may also contain gold.

Because faulted areas are highly mineralized, they display different colors which indicate valuable mineral ores, compounds, and oxides. A solid, unfaulted area of rock such as granite seldom produces valuable ores, and *never* gold.

Red, Green, Black, and Heavy

As the old timers put it: "Look for rocks that are red, green, black, and heavy." These men had a practical knowledge of geology learned

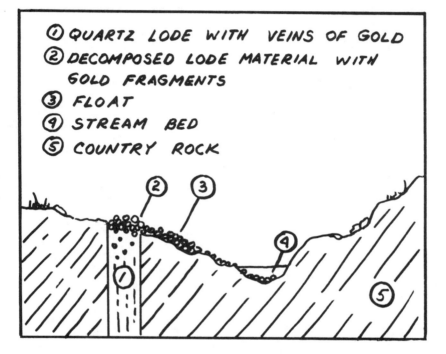

1. QUARTZ LODE WITH VEINS OF GOLD
2. DECOMPOSED LODE MATERIAL WITH GOLD FRAGMENTS
3. FLOAT
4. STREAM BED
5. COUNTRY ROCK

RESIDUAL PLACER FROM DECOMPOSED QUARTZ LODE

Where the nuggets come from.

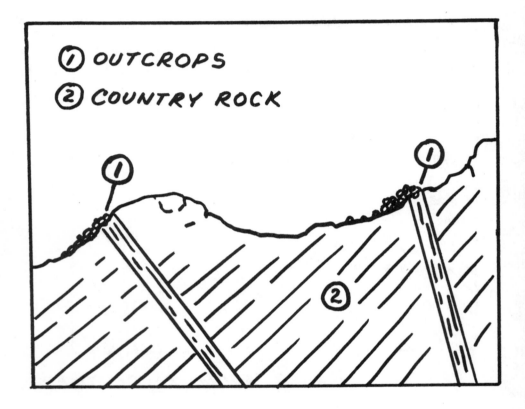

An outcrop is the edge or surface of a mineral deposit or sedimentary bed which appears on the surface.

from long experience. They had found that certain mineral compounds, including iron, manganese, copper, galena (lead sulfide), and silver were somehow associated with gold. Black sand always looked good. Red characterizes many of the iron compounds, green and blue indicate copper, manganese ores are black, galena is a light gray, and these heavy ores frequently contain both gold and platinum. This explains the geological logic of the expression.

Another old saying of the early prospectors was that "You can find iron without gold, but can't find gold without iron." These old timers believed that iron was the "mother metal," as they expressed it. It may be that they were right, because few valuable minerals are found unless iron is present in the same rock formation.

Outcrops and Floats

Veins are often visible at the surface and are spoken of as "outcrops." A mineralized vein will often be partially or entirely covered by debris and surface material. A prospector will sometimes find a piece of good ore out loose on the surface of the ground or partly covered by debris. A piece of ore not part of a solid rock formation was called a "float" by the old prospectors. A float is like a footprint. It is evidence that a vein is somewhere in the area. Since small particles of gold, called "colors," may have been released from other disintegrated pieces of float, the prospector can take samples on the uphill side away from the float and wash them in his gold pan. The trail could lead to a hidden formation or ledge. When samples no longer produce gold, the prospector can conclude that he is above the vein. Trench from the highest point where the gold particles were last found and continue on up the hill until the mother vein is uncovered. The gold pan can be of great help to the prospector in this way.

How Gold Was "Made"

The old prospectors discovered that sulfides, such as silver sulfide or copper sulfide, were often found in the same rock formations as gold. Quite often, very fine particles of free gold are mixed with these ores. The combination is physical, not chemical. Since it is free gold it can be recovered by smelting. They thought that iron pyrite, or fool's gold, was a young form of the sulfide ores which appeared to produce gold when processed. Consequently, they reasoned that the sulfide ores were a further stage in the formation of pure gold. What they did not realize was that the true source of the gold was the finely divided, but free, gold mixed with the sulfide ores, rather than the sulfide ores themselves.

The prospector's theory was that gold was the end product of a long series of continuous changes brought about by various forces and processes. Because gold is associated with certain minerals (particularly iron) the theory held that iron, sulfides, and quartz were necessary for gold to be brought into existence. Observation and constant checking of their theories against the contents of their gold pans convinced the old timers that this assortment of rock and minerals had to be present for gold to "make" as they called it.

This interesting bit of the old prospector's logic and theory probably wouldn't be accepted by modern day geologists, but who can prove they were wrong?

Develop a Prospect—Follow the Lead

When the prospector was fortunate enough to make a discovery his first concern was to "stake" his claim, thus establishing his right to the mineral. Details will be given in a later section on how to stake and file on a claim. Having staked his claim, his next concern was to develop it. To develop a prospect, it will first have to be determined if the claim has values sufficient to make mining it economically feasible. Samples must be obtained and assays taken of the rock formations. As you sink a shaft or work a drift, increased gold in the assays indicates that you have what is called a "lead." One of the maxims of the old timers was to "follow your lead," and this is still good advice today. As long as the values continue to increase, the lead should guide you to a pay streak of rich ore. There should be gold in the pan after the representative ore sample is washed. When the pay streak becomes rich enough to mill or ship to a smelter, you are in business. It is just a question of getting the ore out of the ground, up to the mill, and through the processing system.

A lightweight portable drill is essential in obtaining rock samples for assay purposes. The custom built drill shown here is mounted on a two wheel trailer. It produces a pulverized, dry rock sample and delivers it to a plastic bag four feet long by four inches in diameter; representing the cuttings from ten feet of a $2\frac{3}{4}$ inch diameter hole. The profile of the formations can be studied from these bags and samples taken from any portion with a probe.

Test Drilling and Assaying

To obtain samples of the formations on your claims for assaying, test drilling is necessary. If ground water is present in the formations it will probably be necessary to have conventional hard core samples taken by a drilling contractor. If the formations are dry, then a "dust" core, which is less expensive, may be taken. This consists of the cuttings from the bit as it drills down through the rock. These cuttings are blown and sucked up out of the hole and caught in a four-foot long by four-inch diameter plastic bag, which represents a ten foot section of the rock. The holes are drilled in a gridiron pattern, or on sites selected by a mining engineer or geologist. From the assays the location of the ore body and the value of the ore per ton can be determined. If sufficient drilling is done the total tonnage and value of the ore body can be determined.

From the cores obtained, samples are taken of only those portions that show promise, and assays are then made. Obtaining accurate assays is essential. Large operators often have their own assaying facilities, operated by company chemists. Small miners and prospectors must rely on the services of commercial assay firms. A sensible prospector will have his assaying done by a firm recognized for the high quality and accuracy of their assay reports. A potential purchaser of a mining property will seldom inspect it or consider buying unless he has some evidence of value as determined by assay reports. Also, try to have your assaying done by a firm approved by your potential purchaser. A possible buyer often requires that the assay samples be obtained by an impartial operator such as a core or test drilling company. Finally, a gold prospector should not overlook the possibility that an assay report might disclose that his claim contains minerals that might be of more value than gold!

Develop Reserves

Another good rule is to develop a known reserve of ore—one ton in reserve for every ton mined out. Often the total effort is directed toward mining the known ore body, giving no thought to finding more

P. M. CRISMON, PRES.

CRISMON & NICHOLS

ASSAYERS AND CHEMISTS

440 SOUTH 500 WEST ST.

PHONE 363-7417

P.O. BOX 1708

REPORT OF ASSAY

BALT LAKE CITY, UTAH 84110 — March 19, 1974

WE HAVE ASSAYED YOUR _____ four _____ SAMPLES AND FIND _____ them _____ TO CONTAIN AS FOLLOWS:

DESCRIPTION	NO.	OZS. GOLD PER TON	OZS. SILVER PER TON	PER CENT LEAD	PER CENT COPPER	PER CENT ZINC	PER CENT INSOL.	PER CENT IRON	PER CENT	VALUE OF GOLD PER TON
Silver Key	1		None							
-"-	2		0.10							
-"-	3		Trace							
-"-	4		Trace		0.08					

REMARKS:

CHARGES $ _____ 21.00 _____

CRISMON & NICHOLS

BY

Assay report on Silver Key Claims No. 1, 2, 3, and 4, near the Salton Sea in California.

ore. If your lead plays out it can mean financial disaster if a new source of ore has not been found to replace the worked-out vein. Core drilling beyond the vein will indicate the presence of additional ore. If there is none, and the firm expects a continuous operation, new properties must be found. For more detailed information on the development of a mine prospect refer to the Montana Bureau of Mines and Geology, Miscellaneous Contribution No. 13, *Practical Guide For Prospectors in Montana.*

Chapter VI
Powder—Fire in the Hole!

Sooner or later, the old-time miner would find he could not go farther in working his vein without doing some blasting. This was when he learned to handle powder. In those days, black gunpowder (made of charcoal, saltpeter, and sulfur) was used for blasting. The black, granular powder was poured into a hole that the prospector had laboriously drilled into the hard rock with his long, chisel-like bit, heavy hammer, and strong arm. The fuse, which was in effect a small tube filled with gunpowder, was strung out as far away from the hole as required for safety. When lit, the burning powder in the tube slowly burned its way to the other end and ignited the charge. The explosion would shatter the rock so the miner could dig on for a few more feet.

Dynamite

Although dynamite was discovered in 1866 and replaced powder for blasting purposes, the term "powder" is still used when referring to dynamite. Dynamite is an explosive mixture of glycerin, sodium or ammonium nitrate, and a filler of combustible pulp, such as wood meal. Although it is less sensitive to shock than nitroglycerin, it will still detonate if subjected to a sudden strong impact. Before you even *think* about buying dynamite, get a copy of the *Blaster's Handbook*, published by E. I. du Pont de Nemours & Co., Wilmington, Delaware. I'm not going to give you all the details you will need to know in working with dynamite, since you can get these from the handbook, but let me warn you that dynamite *is dangerous* and that this is one place where excuses don't count and where there is no second chance. You do it right the first time or you wind up doing your prospecting in the Great Mother Lode in the sky. However, there are a few things which are not in the handbook that I have learned from my own experiences; the life they save may be yours. For instance, anyone using dynamite should know that tamping with a steel rod is inviting disaster. Tamping means driving or pounding dirt or sand down into a drill hole above the charge of powder or dynamite. Almost everyone knows that steel striking a hard, flint-like rock will create a spark. A spark is one thing you *don't* want near that powder. Use wood and tamp lightly! Fresh dynamite can take about a fifty pound tap with a wooden stick.

Old Dynamite Is Dangerous

Dynamite left for a year or more has a tendency to form pockets or bubbles of nitroglycerin. Freezing will cause this too. Aging and freezing are the two things to watch out for with dynamite. If little white crystals have formed on the ends of the stick, handle them just as though they were pure nitro. The best thing to do is to take the sticks to the nearest open field, pour gasoline on them and then lay a cap with the proper wires attached alongside of them. Stand well back and set them off. They won't make a loud noise—just a flash of flame—and that is the end of your worry. Old dynamite is very dangerous! *DON'T try to use it! DESTROY IT!*

Setting a Charge

To set a charge of fresh dynamite, put as many sticks in your hole or holes as the hardness of the rock requires. Your handbook will explain this. The last stick to go into the hole has the cap, either an electrical cap or a cap and fuse. Use a wooden stick about the size of the diameter of the fuse to bore a hole about two inches deep into the dynamite, then insert the cap. Loop the wires of the fuse around the stick of dynamite, so the cap will not accidentally be pulled out if someone trips over the wires. Stand clear; hide behind a tree or a large rock, or far enough away to avoid the possibility of rocks falling on your head. Then set off the charge. A flashlight battery is enough to set off a single cap, but use one that will provide at least a six volt charge. Allow at least two volts for each additional cap, to be sure of being very safe. Be alert! A charge that does *not* detonate can still be *very* dangerous. Some men, who didn't live to tell about it, have drilled right into an area where an old charge was located, thinking it had exploded or was free from dynamite.

A Delayed Blast

Dynamite can be deadly in peculiar ways, especially if you are careless. It almost happened to me many years ago. I was working a mine near La Porte, California; it was medium hard rock, and to put a shaft down it was necessary to use dynamite. I was down about fourteen feet when I ran into a situation that almost caused me to consider throwing away the booklet I had on how to safely handle dynamite.

I was using a fuse and caps to set it off. I put a small charge of about fourteen sticks into the hole and then crimped the cap onto the fuse with a crimper, which is a little tool similar to a pair of pliers that crimps one end of the cap around the end of the fuse. The purpose is to hold the cap onto the fuse. I lit the fuse, got out of the hole, headed for cover, and waited for the blast to come. It didn't come. I waited about five minutes more. Something told me not to go near the hole. It was about 11:15 A.M., so I decided to eat an early lunch and give it time to go out completely. Then I planned to put a fresh stick in alongside the rest with a new fuse and cap and set it off. All fifteen sticks in the hole would then detonate.

I took my time about eating. After lunch, I started back up towards the diggings from my camp. It was about 12:10 P.M., and the shaft was approximately 1000 feet from my camp. I was walking along with the fresh stick of dynamite and the fresh fuse and cap in my hand—about twenty to twenty-five feet from the hole—when the charge blew! I hit the dirt and small pieces of rock fell all around me; some of the smaller pieces hit my back and legs. It didn't hurt me, but it sure shook me up: I suddenly realized that if I had started from camp one minute sooner, I wouldn't be here to tell about it.

Needless to say, I didn't do much more that day. I decided it wasn't my day to set off dynamite. I sat down on a log and tried to figure out what had happened. I had heard old timers tell stories about such things, but I didn't think it would ever happen to me. I finally decided that this is what had happened. The crimp was too tight on the fuse. It prevented the powder or fire from reaching the cap. The outside of the fuse sat there and smouldered for nearly an hour. The fire finally reached the powder portion beyond the crimp, which caught fire, and finally detonated the charge.

Tape the Cap

I sat there on that log until I figured out a method to avoid any possible reoccurrence of this accident. After several minutes, I realized that the solution was to *not* crimp the cap! Then I asked myself how I could keep the fuse from coming out of the cap if it were not crimped. The thought came to me: why not *tape* the fuse onto the cap? I had some electric tape with me, so I tried it. I put the fuse into the cap and then taped it together; that way, the fuse could not come out of the cap. I tried setting it off, using just the fuse and cap without the dynamite. It worked. I then realized that this would also seal out any water—such as seepage from the drilled hole—that might make the charge fail to detonate. I tried leaving the cap and fuse to soak in water for about an hour before I lit the fuse: it still exploded. Using this method, I have never had any more trouble. I understand that the handbooks still recommend the use of crimpers to attach the fuse to the stick of dynamite, but I have a good reason to believe that my method is the better one.

Premature Blast

Several years later, I was working in the same general area again. At this time, a construction crew was working on the top dam of the Feather River Project, which is about three or four miles from La Porte in the Little Grass Valley. While I was in town visiting a friend, I heard several ambulances scream by, going out to the dam. About thirty minutes later they came screaming back again, heading for the hospital at Oroville. Still later, some of the workmen came into town and told me that a charge of dynamite was set off accidentally by a sudden lightning storm. The lightning bolt had struck near the detonator box, where the wires leading to the dynamite charge were strung out. The crew was working on the charge when the bolt struck. Unfortunately, several men were killed and the others were seriously hurt. As a safety precaution, one of the wires had even been disconnected from the box and was lying on the ground—in case someone accidentally pushed the plunger on the detonator box.

This event was a great shock to me, as it was to most of the people in that small town. I had trouble sleeping that night. I tried to figure out how this terrible accident had happened and how such a thing could be avoided in the future. To some extent this was a personal problem for me; I had been using electric caps to set off my dynamite for several years. I found them to be more convenient than a cap and fuse, as I could decide to the split second when the charge was to be set off—a simple matter of pushing the handle of the detonator.

Lead Seal

After a time, I realized that there was a simple solution to the problem. Everyone who handles dynamite and electric caps should know that the caps come from the manufacturer with a small lead seal attached across the two wires that run to the powder part of the cap. This lead seal shorts out the wires, making it impossible for the cap to detonate. Many good powder men don't seem to realize that this lead seal should *never* be broken until the wires are all strung out and the

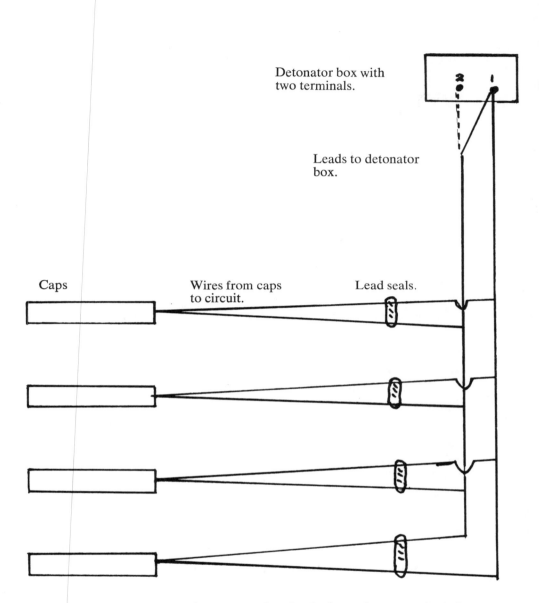

Detonator box with two terminals.

Leads to detonator box.

Caps

Wires from caps to circuit.

Lead seals.

How to prevent premature detonation. This sketch shows the circuit from the detonator box to four electric caps. The lead seal across the two wires of each cap creates a dead short which prevents premature detonation. These seals should not be broken until you are ready to fire the charge.

The leads to the detonator box should both be attached to the No. 1 terminal until ready for firing, then one wire should be removed and attached to the No. 2 terminal. This creates an open circuit ready for firing.

entire charge ready to go. The too-early breaking of this seal has been a primary cause in many seemingly mysterious dynamite accidents. In stringing out the wire from the detonator box to the site of the charge, it is common practice—as a safety measure—to drop one of the two wires to the ground instead of connecting it to the terminal on the box, just as was done by the powder man at the dam accident site. Of course, the wire would be attached to the terminal just before the operator was ready to set off the blast.

Short the Circuit

In analyzing what had happened in the lightning accident, I concluded that the lightning bolt could not have completed the circuit to the electric cap if the operator had temporarily placed both wires on one terminal, instead of dropping one wire on the ground. This would have created a dead short, capable of blocking any flow of electricity to the caps.

In practice, the only absolutely safe procedure is to keep the two wires connected to *one* terminal of the detonator box, and to keep the seals on the electric caps *unbroken* until all the wires are strung out and the charge is ready to go. Then, when the operator is sure that all the men are clear, he can send one man to the charge to remove the seals from the electric caps. When this man is clear, the operator can take the second wire from the first terminal and place it on the second terminal, thus opening the circuit to the electric caps. The entire system is now ready. All that remains is for the operator to shout "Fire in the hole," and push down the plunger. If this very simple routine is followed, the possibility of an accident is reduced to practically zero, even if lightning does strike nearby as it did in the dam accident.

Transmitter Can't Cause Blast

Most people traveling by automobile through areas where road construction and blasting are going on have noticed signs stating

"Turn off Transmitter—Blasting." Such signs would be entirely unnecessary if the procedures outlined above were followed; the use of these signs indicates that either the man in charge doesn't understand the basic principles explained above, or that he is compelled to put up the signs by some law or regulation that is a result of ignorance. There is no way that a radio transmitter can cause a premature detonation when the electric cap seals are unbroken and the lead wires are shorted out by placing them both on a single terminal of the detonator box.

If you are a weekend gold panner, steer clear of dynamite: you don't need it. If you are serious about hard rock mining, *be careful!* If you follow the precautions above you shouldn't get in serious trouble with dynamite.

Two Component Explosive

Recently, an entirely new form of explosive has come into the picture. It utilizes two individual components—a liquid and a powder—which can be stored or transported as non-explosives until they are mixed on the site. This is unique because neither of the components alone is explosive, but when mixed together they become as powerful an explosive as dynamite. The explosive is mixed and used in two forms: as a stick and in a pouch. When the dry powder in the stick or plastic pouch is armed by pouring in the liquid it can be readily ignited with either electric or fuse caps. This new explosive was developed by the Atlas Powder Company. Additional information may be obtained by writing this company at P. O. Box 2354, Wilmington, Delaware 19899.

I was informed by another source that the idea for this type of explosive was the result of a tremendous accidental explosion which occurred on the Gulf Coast several years ago. A ship loaded with ammonium nitrate fertilizer blew up at the dock, devastating the area and killing a large number of people. Subsequent investigation disclosed that fuel oil from the storage tanks for the diesel engines of the ship may have leaked into the cargo of fertilizer. Experiments with this type of mixture disclosed it to be highly explosive when ignited.

We will never know what ignited the explosive mixture, but out of this terrible accident came the idea for a new form of explosive.

In open pit mining and other large industrial applications, a mixture of ammonium nitrate fertilizer and diesel fuel is now replacing dynamite as a cheap, effective explosive. This development probably had the same origin.

Chapter VII
How to Stake a Mining Claim

Before discussing how to stake a claim and what that involves, it should be clearly understood that we are only talking about Federally owned land. These areas are known as public domain or public lands and are administered by the U. S. Forest Service under the Department of Agriculture, or by the Bureau of Land Management under the Department of the Interior. Under Federal regulations, it is only on these public lands that you may prospect for minerals and stake a claim. All lands in the United States not owned by the Federal Government are privately owned and are not open to prospecting unless arrangements are made with the owners.

What Is a Mining Claim?

Briefly, an unpatented mining claim is an area of land on which an individual has obtained a right to extract and remove minerals by the act of valid location under the mining laws of 1872, but where full title has not been acquired from the United States Government. This is known as a possessory right, and is maintained by the performance of certain prescribed annual work and recording requirements. This right may be sold, inherited, or taxed according to state law.

Mining claims were officially brought into existence by an Act of

91

Congress in 1872, called "An Act to Promote the Development of Mining Resources of the United States." The complete law is found in the United States Code, Title 30, Sections 2k-54. The regulations can be found in the Code of Federal Regulations (CFR), Title 43, parts 3400-3600, which are available from any Bureau of Land Management office.

Mining Districts

Prior to the enactment of these basic mining laws of 1872, the regulation of mining activities was done through "Mining Districts." After the discovery of gold became known and people arrived in numbers it became evident that some form of government had to be organized to control mining activity and to protect the rights of the individual miners. Meetings were called by the miners who owned claims staked on a certain creek or land area. At these open meetings, which were conducted by a leader selected by those present, rules were set down to establish and protect the property rights of the miners. Each land area was defined and named as a particular mining district. As other problems related to keeping the peace, establishing order, and administering these rules arose, mining districts functioned as government until the organization of the Territories. A central place was designated where the claim boundaries and their dates of delineation were recorded. This was generally a designated building in the largest settlement or town of the mining district which then became the seat of government and was the prototype for our present day counties and courthouses.

When formal government was established in the territories and states, the original need for mining districts disappeared. Today, they are of historical interest only, although the name given to each district still identifies specific areas within the political subdivision of the county and state. The forms used to record mining claims, called "notices of location," still require the name of the mining district in which the claim is located. In addition, the notice requires a description of the location of the mining site (called the "discovery") in relation to some permanent and natural object within a stated county, range, township, and section. This ties the old mining district designations to the currently established system of land survey and the legal description of land areas, as well as to the county subdivision of government. Location notices of claims are recorded in the county courthouse as a permanent, public record.

Discovery of a Valuable Mineral Deposit

To be valid, a mining claim may be staked and held only after the discovery of a valuable mineral deposit. The courts have established what is called the "prudent man" rule to determine or define what constitutes a valuable mineral deposit. The discovery is declared valuable if a person of ordinary common sense (prudence) would be justified in a further expenditure of his labor and means to develop this mine, with a reasonable prospect of success. The basis for the prudent man rule and the authority for all federal mining law comes from the CFR, Title 43, Code mentioned above.

State Mining Laws

Each state has individual mining laws, so there are various differences in the requirements from state to state. These laws parallel the Federal Code, however, and do not conflict with it. The laws of the state in which you wish to stake your claims may be secured from your State Bureau of Mines or your local public library.

Staking the Lode Claim

Federal law requires that a claim be marked distinctly enough so that it can be readily identified. Each state has expanded this general requirement to include detailed directions for marking claim boundaries. The minimum requirements are listed as follows:

1. Set a stake at the point of discovery and one on each corner of the claim. A center stake may be required on each side. Some states require the name of the claim to be painted on each stake, with a designation on each stake regarding center, or direction; i.e., Northwest corner, Southwest corner, Center, etc.

2. The location notice, giving the name of the locators, the name of the claim, whether lode or placer, the description of the location,

Staking the claim.

any minerals claimed, and meeting any other requirements made by the state, must be placed at the point of discovery. This should be placed in a waterproof container, plainly visible beside the discovery monument or stake.

3. An exact copy of the location notice must be filed at the county Clerk and Recorder's office in the county in which the claim is located.

4. A lode claim is limited in size by statute to a maximum of 1500 feet in length and 300 feet on both sides of the discovery lode or 600 feet total width. The longer sides are parallel to the course of the vein. The location is described by metes and bounds, a term meaning that the direction in degrees and distance in feet is measured from point to point and return to the place or point of beginning.

5. If the location description of either a lode or placer claim is not tied in with the legal subdivision system, it must provide the distance and direction as accurately as possible from the discovery point to some well-known reference point, such as a point on a steel bridge, a fork in a stream, a road intersection, or some other known landmark within the county.

Diagram of the Sweetgrass Lode claim.

R	*Reference point (corner of steel bridge)*
D	*Discovery of valuable mineral*
XY	*Course (or direction) of vein*
AB	*Northeast boundary line of claim, 1500 feet long, parallel to course of vein*
CS	*Southwest boundary line of claim, 1500 feet long, parallel to course of vein*
BS	*Southeast end of claim, 600 feet wide*
AC	*Northwest end of claim, 600 feet wide*
NB and MA	*Distance from center of vein to northeast boundary line—300 feet.*
NS and MC	*Distance from center of vein to northwest boundary line—300 feet.*

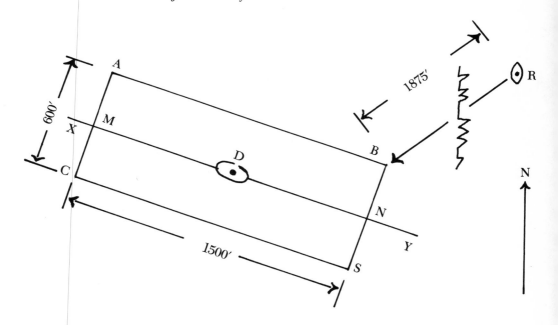

Metes and Bounds Land Descriptions

Metes are measures of length in feet and bounds are boundaries. A metes and bounds description locates the land area with reference to permanent objects or readily distinguished landmarks in the vicinity of the land area to be described.

Following is a description of how the various points on the diagram are determined:

1. The line XY is determined by the point of discovery "D" and the course or direction of the vein.

2. Points M and N are determined by measuring 750 feet each way from point D along the course of the vein or line XY and including point D, thus establishing the length of the claim at 1500 feet, the maximum length by statute for a lode claim.

3. From point N measure 300 feet on a perpendicular or a 90 degree angle from XY. This establishes point "B."

4. In like fashion, points S, A, and C are established, thus determining the two boundary lines AB and CS parallel to the course of the vein XY.

5. Points A and C, and B and S, when connected form lines which comprises the two ends, 600 feet in width, parallel to each other—the maximum width by statute for a lode claim.

6. The next step is to find some point of reference—in other words, some permanent and recognizable place to tie to. We select the northwest corner of a steel bridge over Ruby River, 4 miles south of Twin Bridges, Montana, as this point of reference which we designate as point "R."

7. With the aid of a compass, we determine the angle from point R to point B and measure the distance from R to B, which we find to be 1875 feet. We now know the location of point B with reference to point R.

Describing Your Claim

For the purpose of preparing a description to identify the location of your claim on the location notice, you can estimate the approximate direction from your reference point on the corner of steel bridge as simply "Southwest." By stepping off the distance you find it is 625 steps of three feet each, giving 1875 feet to the nearest corner "B." Now, you can state the location of your claim on the location notice as follows:

From the southwest corner of the Roscoe Bridge on the Ruby River, Madison County, Montana; approximately 1875 feet in a southwesterly

Surveying the claim.

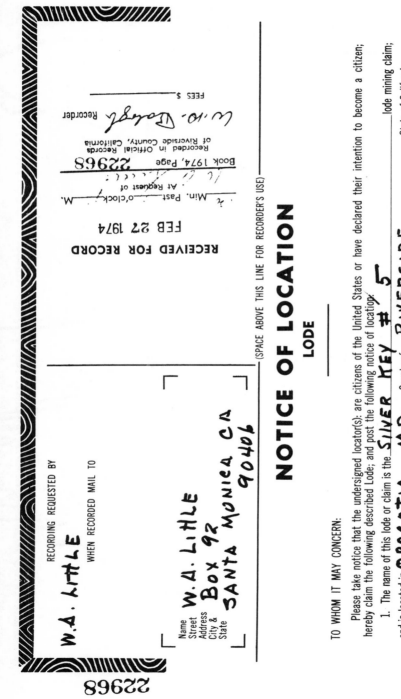

(SPACE ABOVE THIS LINE FOR RECORDER'S USE)

NOTICE OF LOCATION
LODE

TO WHOM IT MAY CONCERN:

Please take notice that the undersigned locator(s): are citizens of the United States or have declared their intention to become a citizen; hereby claim the following described Lode; and post the following notice of location:

1. The name of this lode or claim is the **SILVER KEY # 5** _____ lode mining claim; and is located in **OROCOTIA M.D.**, County of **RIVERSIDE**, State of California.

2. The name, current mailing address or current residence address of the locator(s) are given at the bottom of this notice.

3. The number of linear feet claimed in length along the course of the vein each way from the point of discovery is **750** feet in a **N.W.** direction. (Not to exceed 1500 feet)
   feet in a **S.E.** direction, and **750** feet in a _____ direction. (Not to exceed 1500 feet)

4. The width on each side of the center of the claim is **300** feet. (Not to exceed 300 feet)

5. The general course of the vein or lode is _____

6. The date of the posting of the original of this Notice, which is the date of location, is **FEB 22,** 19**74**.

7. This claim, described by reference to some natural object or permanent monument as will identify the claim located, is as follows:
   SEC. 20 T6S, R11E SBM- 50 ± HWY 195

of this claim, together with all water and timber and any other rights appurtenant, as allowed by the laws of this State and of the United States.

LOCATORS

NAME   STREET ADDRESS   CITY, STATE and ZIP CODE

1. W. A. LITTLE   203 E. 4ᵗʰ ST. BOLDER, Mont. 59632

2. VERNE H. BALLANTYNE   217 W. Koch St. Bozeman, Mont. 59715

3.

4.

STATEMENT OF THE MARKING OF THE BOUNDARIES
AND OF PERFORMANCE OF DISCOVERY WORK

NOTICE IS HEREBY GIVEN by the undersigned locator(s) that in accordance with the requirements of the California Public Resources Code:

1. The above notice of location is a true copy of said notice; and is hereby incorporated by reference herein and made a part hereof.

2. The locator(s), within the following time, as required by law, have defined the boundaries of this claim by erecting at each corner of the claim and at the center of each end line, or the nearest assessable points thereto, a conspicuous and substantial monument; and each corner monument so erected bears or contains markings sufficient to appropriately designate the corner of the mining claim to which it pertains and the

name of the claim. The date of marking is: _____ ; and the description of monuments are:

_____

_____

3. The United States survey within which all or any part of the claim is located is: Township _____ ,

Range _____ and Meridian _____ .

4. The locator(s) have performed the following location work: _____

_____

_____

DATED: _____ , 19 _____   LOCATOR(S) _William A. Little_

_____

_____

_____

SEE REVERSE SIDE

NOTICE OF LOCATION—LODE—WOLCOTTS FORM NO. 1134—REV. 3-71

DO NOT RECORD

1. This form should not be used where a placer location notice, tunnel site location notice, or mill site location notice is required.

2. Section 2316, Public Resources Code, reads as follows:

"(a) A wooden post or stone structure not less than 3½ inches in diameter or a metal post not less than two inches in diameter, projecting at least three feet above the ground and set at least one foot in the ground, or a mound of stone at least three feet in height above the ground, is prima facie a conspicuous and substantial monument as referred to in this chapter.

(b) Where by reason of precipitous ground, it is impractical or dangerous to place a monument in its true position, a witness monument may be erected as near thereto as the nature of the ground will permit and marked so as to identify the true position.

(c) Where by reason of working the claim, it is impractical or dangerous to maintain a monument in its true position, a witness monument shall be erected as near thereto as the nature of the ground will permit and marked so as to identify the true position."

3. Within 90 days after the posting of a Notice of Location, the locator(s) " . . . shall record in the office of the county recorder of the county in which such claim is situated a true copy of the notice together with a statement by the locator of the markings of the boundaries, and of the performance of the location work, as well as the character of each, which statement also shall include the township, range and meridian of the United States survey within which all or any part of the claim is located."

Section 2312, Public Resources Code.

This standard form covers most usual problems in the field indicated. Before you sign, read it, fill in all blanks, and make changes proper to your transaction. Consult a lawyer if you doubt the form's fitness for your purpose.

This is a standard NOTICE OF LOCATION (lode) form. Various printing establishments make these forms and they are sold in stationery stores wherever there is a need for them.

The lower part of this form was not completed as the discovery work was not yet done at the time of filing. Assay reports from samples taken from an outcrop at the time of filing were negative so the discovery work was never done and the claim lapsed.

direction to point "B," the northwest corner of this claim; from point B 1500 feet northwest, parallel to the trend of the discovery vein "D," to point "A," thence 90 degrees southwest 600 feet to point "C," thence southeast 1500 feet, parallel to the trend line, to point "S," thence northeast 600 feet to point "B," the point of beginning.

This improvised description, while not entirely accurate, is sufficient and legal as far as staking your claim is concerned. (Note: This is not an actual location.)

Surveying the Claim

Neither the Federal nor the state laws require that mining claim location descriptions be established by an official surveyor. However, if you find that you have valuable minerals and start to mine or prepare your claims for sale, it is a wise policy to have a survey made by a licensed surveyor and to have this more accurate description recorded in the county Clerk and Recorder's office.

A description prepared by a qualified surveyor of your claim would read like this:

Beginning at point "R" (the southwest corner of the Roscoe Bridge on Ruby River), thence south 79 degrees—15 minutes on a true bearing a distance of 1875 feet to point "B," the true point of beginning, thence south 18 degrees—45 minutes west, a distance of 600 feet to point "S," thence north 71 degrees—15 minutes west a distance of 1500 feet to point "C," thence north 18 degrees—45 minutes east 600 feet to point "A," thence south 71 degrees—15 minutes east a distance of 1500 feet to point "B" the true point of beginning, lying in SW¼SE¼ Sec. 30 T14N, R23W.

Discovery Work

Part of the requirement to establish a valid claim is the performance of certain discovery work. State mining laws vary somewhat in their requirements, but a ten foot deep pit or shaft is generally required. Drilling a twenty-five foot test hole can usually be substituted for the pit or shaft. Ordinarily, a certain period of time is allowed from the time the claim is staked or filed until the required discovery work must be completed. In some states, this period is ninety days. You must secure additional information to be sure that you are complying with the laws in the state where you are staking your claim.

Placer Claims

Placer claims are limited in size to 20 acres per claim per locator. An association of two or more persons, up to a total of eight, may locate 20 acres per locator, making a total or maximum of 160 acres. Where practicable, placer claims are located by legal subdivision. A legal subdivision is a section or a part of a section lying in a designated township, range, and principal meridian, for instance, S½SE¼ Sec. 14, T6N, R9E, MPR. On unsurveyed land or where not practical because of the terrain, placer claims may be located by metes and bounds.

Assessment Work

A minimum of $100 worth of work must be performed each year on, or for the benefit of, your claim in order to maintain your right to this claim. This work ordinarily consists of extending the depth of the shaft, the length of the tunnel, or doing other work on, or for, the benefit of the claim. This work must be done sometime during the mining year (September 1st to September 1st) and must be completed on or before noon on September 1st of each year. A statement that this work has been performed must be recorded in the county recorder's office within a specified period after September 1st. Again, state requirements are different. Be sure you have the right information for your state. This work requirement is known as annual assessment work.

Patenting the Claim

The rights of the owner of a mineral claim are limited to possession for the purpose of developing and extracting discovered valuable minerals. To obtain the additional rights included in full ownership, a patent must be obtained. This can be quite expensive, as there are many legal requirements that must be satisfied before a patent can be issued. Among these requirements are:

1. Proving a valuable discovery.
2. Survey by cadastral engineer.
3. File a plat of claim.
4. Post 60-day notice on claim.
5. Apply for patent—$25.
6. Not less than $500 work.
7. Full description of vein.

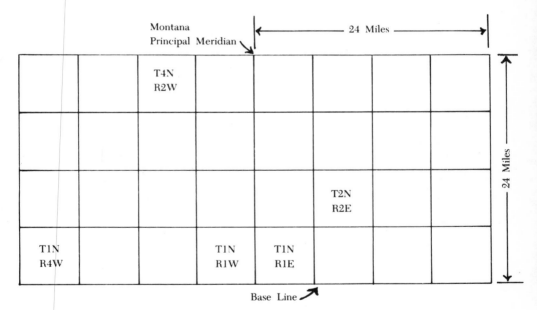

U.S. Governmental Survey System—The Tract. The largest squares measure 24 miles on each side and are called checks or tracts. Each tract is further divided into 16 squares called townships whose four boundaries each measure six miles. The columns of townships running north and south are called ranges and are numbered according to their distance from the principal meridian (the main north-south survey line).

U.S. Governmental Survey System—The Township. A township is six miles square or 36 square miles. Each square mile is designated as a section which is equivalent to 640 acres.

The rows of townships running east and west are numbered according to their distance from the baseline (the main east-west survey line).

Sections within a township are numbered from the northeast corner, following a back-and-forth course until the last section in the southeast corner is reached. For purposes of land description, sections are commonly divided into half-sections containing 320 acres, or quarter-sections containing 160 acres. Land descriptions are made by referring to a particular quarter of a particular section located within a particular township, county, and state.

U.S. Governmental Survey System—The Section. The 40 acres in the southeast corner of the Section 2 above, located in Township 5 North and Range 6 East would be legally described thus:

SE ¼ of the SE ¼ Section 2, Township 5 North and Range 6 East of the Montana Principal Meridian, County of Gallatin, State of Montana.

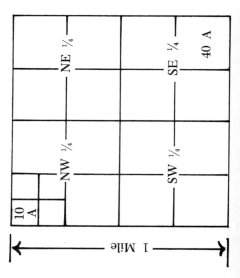

Township grid (T6N / T5N, R 6E / R 7E, 6 Miles):

6	5	4	3	2	1
7	8	9	10	11	12
18	17	16	15	14	13
19	20	21	22	23	24
30	29	28	27	26	25
31	32	33	34	35	36

Section grid (1 Mile):

NW ¼ | NE ¼
SW ¼ | SE ¼

10 A
40 A

8. If placer, prove is placer.
9. Prove possessory rights.
10. Other supporting papers.

From this partial list of requirements, you can see that obtaining a patent is not a simple procedure. Anyone intending to file an application for a patent of their mining claims should contact a Bureau of Land Management or Forest Service office for complete information.

If you are discouraged by this list of requirements, you may be consoled by the fact that a patent to your mining claim is not necessary for carrying on mining activity, as long as the legal requirements to maintain your right as a claimant are performed. These requirements are 1. being able to show the discovery of a valuable mineral, and 2. performing a minimum of $100 of annual assessment work and recording this in the Clerk and Recorder's office in the courthouse of the county in which the claim is located.

For more complete information on how to obtain a patent to a mining claim you are referred to a small pamphlet available from the Bureau of Land Management, U. S. Department of the Interior, entitled *Patenting a Mining Claim on Federal Lands—Basic Procedure.* This is a free publication available from any of the twelve Land Offices of the Bureau of Land Management. For Montana and North and South Dakota the address is Montana Land Office, 316 North 26th St., Billings, Montana 59101. For Arizona it is Arizona Land Office, Federal Building, Phoenix, Arizona 85025.

The Small Hard Rock Miner

The hard rock miner can be "in business" on practically any scale, from small to large, although in general rock mining requires more original investment and more knowledge of mining than is required of the placer miner. It is safe to say that the ore from every mine has its own individual characteristics and requires a method of handling and processing specifically related to those characteristics. To go any further into the methods of handling or processing ore is beyond the scope of this book.

However, the following two pictures, and the accompanying description, are included to give a general idea of one rather typical small hard rock miner's operation from the handling and processing point of view. This miner was able to secure a second hand crusher at a reasonable price, and since he was a welder and mechanic, built the

This millsite in Josephine County, Oregon, is equipped with an old-style crusher on the left and a high speed dry mill on the right. The ore is crushed to nut size in the crusher, then put in the dry mill where it is reduced to a powder. The dry mill is custom built and similar to a huge fan or water wheel with four vanes or paddles. The ore is mined from a nearby outcrop.

dry mill himself. The shaker table was also a used piece of equipment.

This is definitely a "poor boy" operation as far as the equipment is concerned. However, the most modern and expensive mining and handling machine does not guarantee, or even imply, success in a small mine operation. The knowledge and experience of the miner or the guiding hand and intelligence of his engineer or advisor is the critical element in a successful small mine operation.

From the dry mill, the powdered ore is blown into a cyclone where it is watered down with a spray, then passed onto a shaker table where the waste rock floats off and the heavy concentrate of minerals and metals is shaken into pans at the foot of the table. The slope of the table, the amount of water allowed to flow onto it, and the amount of vibration are all important factors in this operation.

President Ford signs the private ownership of gold bill—August 14, 1974.

Chapter VIII
Ownership and Marketing

Several other factors are important for the amateur prospector and miner to know and are discussed below.

Private Ownership of Gold

During the period of economic stress and depression of the early 1930's, the Gold Reserve Act of 1934 was passed by Congress. This Act arbitrarily required all persons in the United States to exchange their privately owned gold and bullion for so-called lawful paper money. By this act it became illegal for a citizen of the United States to own gold coin or bullion. This law, together with related legislation, regulations and the licensing of gold buyers, coupled with the increasingly unrealistic and arbitrarily fixed price of gold, destroyed the market for gold for the small miner.

Prior to the passage of this act, it was common practice for the gold miners to ship their gold to the nearest United States mint where it would be graded, refined and purchased or minted into coil or bullion for the owners. This free coinage of gold by the United States Treasury provided a ready market for the producers of gold so long as the arbitrarily fixed price was in balance with other segments of the economy. There was no question concerning the private ownership of gold at that time. Ownership, in the usual sense, was assumed and taken for granted.

On August 14, 1974, President Ford signed Public Law 93-373 of the 93rd Congress which repealed the Gold Reserve Act of 1934 and made it legal for any person in the United States to buy, sell, hold or otherwise deal in gold either in this country or abroad as of December 31, 1974. Thus, the right which was taken from United States citizens over 45 years ago was restored. This historic action officially legalized the private ownership of gold.

Market for Gold

With the passage of the Private Ownership of Gold act, gold became a commodity in the same sense that copper, silver, platinum, wheat, corn, lumber, cotton or any other product bought, sold and used in domestic and international trade is a commodity. Upon the passage of the act, the major commodity markets or exchanges of the nation offered to buy or sell gold futures contracts. There is no restriction on the purchase or sale of gold—it may be sold to anyone who wishes to buy. As time goes on and the gold mines of the country again become productive, new marketing channels will be opened up on a domestic level.

Since the repeal of the Gold Reserve Act of 1934, the price of gold has been determined at certain central world markets, such as Zurich, Switzerland and London, England. These markets will continue to be a major factor in the pricing of gold. Unlike other commodities, the historic role of gold as a standard of value to support domestic currencies and regulate international settlements of trade balances places an unknown factor on the pricing of gold in these world markets.

Supply and Demand

Basically these prices are determined by the forces of demand and supply, but behind the demand for gold are powerful economic and political factors. To offer any comment on these factors is beyond the scope of this book. How and where the prevailing price for gold is established may be of interest to the gold prospector but it does not provide him with a ready market he can use, in the same sense the United States Treasury did prior to 1934.

However, there are responsible market channels, or purchasers of raw gold. The Arizona Department of Mineral Resources, Mineral

Building, Fairgrounds, Phoenix, Arizona 85007 has provided a list of possible buyers of gold. Of the 23 buyers on their list here are 4 of the best known:

American Smelting and Refining Company
Ore Department
120 Broadway
New York, New York 10005
Telephone: (212) 732-9500
Contact: G. W. Anderson, Manager

Delta Smelting & Refining Co., Ltd.
2200 Shell Road
Richmond, British Columbia
V6X,2Pl, Canada
Telephone: (604) 273-2771
Contact: David W. Seed, President

Englehart Minerals & Chemicals Corporation
Refining Department
429 Delaney Street
Newark, New Jersey 07105
Telephone: (201) 589-5000, Telex: 138500
Contacts: R. L. Searle, Commercial Manager,
or J. M. Jacobs, Sales Manager

Handy & Harman
850 Third Avenue
New York, New York 10022
Telephone: (212) 752-3400
Contact: W. F. Parry, Manager, Refining Department
Note: The company has a precious metal refinery in El Monte, California

Not listed by the Arizona Department of Mineral Resources is a consistent and long-time advertiser in the California Mining Journal,

the J. and J. Smelting and Refining Corporation 17474, Catalpa St.,
P. O. Box 727, Hesperia, California 92345
Telephone: (714) 244-5642.

This firm will accept gold for refining to be returned to the owner in
bar form, 999.7 fine gold, or they will purchase it at 100 percent
market price. They request that 10 percent of the gross weight of the
ore or concentrate submitted must be gold.

The firm has been in operation since July 10, 1970. They claim to
handle a large percentage of the gold produced in Alaska and suggest
any bank in Alaska as reference. Contact before sending any product,
preferably by phone. The phone service in the area is poor so it may
be necessary to write, in which case the refinery will answer by phone
or letter. In any case prior arrangement must be made before ship-
ping.

Gold seldom occurs in "pure" form, that is, as 1000 fine or 24 Karat
gold. For instance, gold found in the Mother Lode country of California is between 800 and 900 fine. Some Alaska gold carries as high as
60 percent copper which gives it a red color. One of the most common
metals naturally alloyed with gold is silver, which gives it a mellow,
light gold color.

When selling gold it must be remembered the other metals alloyed
with it are considered impurities and you will be paid only for the
recovered ounces of 1000 fine gold. An exception to this is silver
which is separated and purchased at a percentage of its market price.

The separation of the impurities, including silver, copper and other
metals from raw gold is the task of a refinery and must be paid for by
the owner of the gold. Before sending your raw gold to a refinery you
should know their schedule of charges for various quantities etc.

A very special and profitable market exists for gold nuggets in their
natural state. This is the jewelry business. These nuggets, also called
specimens, may be sold for several times the world market price for
gold. Unusual pieces of hard rock gold, such as quartz containing
veins of gold, are also in demand as specimens and will bring pre-
mium prices from collectors and hobbyists, as well as from jewelers,
far in excess of the market value of their gold content. The exact price

of any individual piece of specimen gold is largely determined by the beauty and uniqueness of the specimen and the bargaining power of the individual.

Intrinsic Value

A term that relates to the value of gold is intrinsic value. This means the actual market value of gold contained in any object. For instance, a gold coin which weighs one troy ounce and is 900 fine actually has 90 percent of an ounce of pure gold in it. Based on a market price of $400 per Troy ounce, the intrinsic value of the coin would be $360.

Coarse gold and nuggets can be sold directly to jewelers.

How to Price Your Coarse Gold

The basic price of gold set in the major world markets is the principal factor governing domestic and local sales of placer gold, it is true. But, when you get into coarse gold (anything one grain and over) you have a different problem. To a large extent it is a bargaining situation. You have a product the jewelers are interested in and there is no central market. Following is a rough guide to use in pricing your coarse gold. It is not authentic in the sense it is suggested by a state or other government body, but it represents the composite opinion of knowledgeable persons that have first hand experience in the gold market.

WEIGHT	PRICE
Fine gold—under one grain	Market price less 5 to 10% based on assay. Same for silver content.
Three to 12 grains	Market price based on weight of specimen only (no assay) plus 10% and up
Twelve grains to one Pennyweight	Market price based on weight only plus bonus of 50%
One Pennyweight up to one ounce.	Market price based on weight only plus bonus of 50-75%
Nuggets one oz. or larger	Nuggets have an individual value, price usually arrived at by bargaining. Market plus 100% and more— depending—.

Rough Guide to Pricing Coarse Gold

Numismatic Value

Another term related to gold is numismatic value, which means the total value of a coin, which includes the intrinsic value as well as the value of it as a desirable collector's item. For example, a common date $1.00 coin might have an intrinsic value of $10 but due to its scarcity and desirability as a collector's item it has a numismatic value of $150 to $250 depending on its condition.

Gold Prospector's Association of America

Assisting, in a small way, with the marketing phase of recreational prospecting and mining is a grass-roots organization owned and run by active gold prospectors and miners, the Gold Prospector's Association of America (GPAA).

From a handful of gold prospectors at its founding in October 1969, to over 60,000 members at the beginning of 1982, the GPAA reflects the increasing interest in gold prospecting during the past 13 years and the drive and dynamic personality of its sounder and president, George Massey.

Another reason for its rapid growth is, in the words of the founder, "The sole purpose of the GPAA is to enhance recreational mining and to protect the rights given under the 1872 mining laws, which include the right of any individual, regardless of financial status, to prospect metals on public land that is open to mineral entry."

To further this purpose, Massey and others in the organization keep close tabs on any move made to erode the rights of the small miner. One success they believe they are responsible for is that which maintains that Forest Service regulations provide for prospecting and recreational mining in wilderness areas.

The association publishes its own news magazine, *The Gold Prospector's News,* which carries timely articles on prospecting and recreational mining.

Gold shows are held in various cities in the western states by the GPAA which provides an opportunity for suppliers of prospector's and recreational mining equipment to exhibit and demonstrate their product.

Free lectures on panning for gold, processing the concentrates, dredging, crevice mining, dowsing, sluicing, Federal regulations and other phases of small mining are given at the show, all free to the public.

Members with gold to sell find the shows provide a market for their gold. They ar offered the opportunity to set up their own display and sell their specimen gold to interested patrons at any of the shows.

Gold's Fingerprints

To speak of gold having fingerprints sounds odd it is true, but the term is only a comparative way of saying that any original sample of raw gold can be identified as to its origin, or where it came from. The fact gold is always mixed with other metals or minerals in a variety of proportions gives any individual sample a unique identity directly related to its origin—the place where it was found and mined. An assay report which gives a complete listing and the percentages of all metals, automatically reveals the location from which it came, if there has ever been gold identified from that area before. It is simply a matter of comparing the assay report with reports from the gold producing areas of the world. Computerized recordkeeping has made this a matter even more simple than before. This unique quality of gold is of great value when it comes to tracing and identifying stolen specimens or quantities of gold.

I heard of an interesting example of this where a man set up a small placer operation on a creek that had been worked out a number of years before by a large dredging company. Even though he was obtaining very little gold from the reworked placer gravel he continued to keep the work going for a couple of years.

This man had been the dredgemaster on the dredge when it worked the creek originally. The speculation was the man had high graded the dredge return, then when he found he couldn't sell the gold without its origin being revealed, waited and went back years later and conducted his placer operation as a cover up so he could sell the stolen gold. Gold does have "fingerprints".

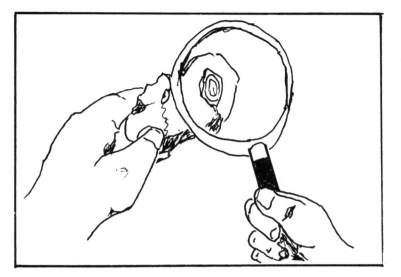

Gold also has fingerprints—any original sample of gold can be identified as to where it came from.

The word NUGGET should be applied only to a piece of water-worn, native gold larger than a grain of weight.

Troy Weight

Gold and other precious metals are weighed by the Troy system. The grain is the smallest unit and reflects the fact that one grain of gold at one time was equal to a grain or kernel of corn, or other cereal. The next larger unit is the pennyweight consisting of 24 grains. The troy ounce contains 20 pennyweight or 480 grains. The pound is 12 ounches.

24 grains	equals 1 pennyweight
20 pennyweight	equals 1 ounce (480 grains)
12 ounces	equals 1 pound (240 pwt & 5760 grains)

Apothecaries System

In the apothecaries of weights, the pound, ounce and grain are the same as the pound, ounce and grain of the Troy system. However, the apothecaries system has other units of weight; the dram is 60 grains and the scruple is 20 grains.

Avoirdupois Weight

The avoirdupois system of weights is used for trade and commercial use in the United States and the pound and ounce are the basic units.

16 ounces equals one pound.

The avoirdupois ounce is *not* the same as the troy ounce, and the avoirdupois pound is *not* the same as the troy pound.

$$
\begin{aligned}
1 \text{ pound av.} &= 16 \text{ oz av.} &= 7,000 \text{ grains troy} \\
1 \text{ oz av.} &= 437.50 \text{ grains troy} \\
1 \text{ lb Troy} &= 12 \text{ oz troy} &= 5,760 \text{ grains troy} \\
1 \text{ oz troy} &= 480 \text{ grains troy}
\end{aligned}
$$

To convert avoirdupois ounces into try ounces, multiply by 0.910. To convert troy ounces into avoirdupois multiply by 1.097. To convert avoirdupois pounds into troy pounds, multiply by 1.215. To convert troy pounds into avoirdupois pounds, multiply by 0.823.

Chapter IX

Regulations

To deal with certain problems arising from the fast-moving gold scene—particularly prospecting for and mining gold—several new Federal regulations have recently come into existence. Information concerning these new regulations, as well as a listing of U.S. Bureau of Land Management offices, is given in this chapter.

United States Forest Service

The Congress of the United States has charged the Secretary of Agriculture with the responsibility for administering certain laws it has passed with regard to the use of the surface of the National Forest System lands. To comply, the Secretary of Agriculture authorized the Forest Service to formulate and enforce new rules and regulations relating to the preservation and use of the surface of the Forest Service lands. These rules were published in the *Federal Register*, Volume 39, Number 168, August 28, 1970, together with a resume of the hearings held prior to their final writing, and became effective on September 1, 1974. Copies of this issue of the *Federal Register* may be obtained from any Forest Service District Office.

However, the regulations in the *Federal Register* are written in legal jargon that is difficult for the layman to understand. To deal with this problem, the Forest Service has published simplified handbooks, available at the Forest Service District Office in your state. The titles are:

- "Mining in National Forests, Current Information Report Number 14," published January 1975.

- "Questions and Answers about the 1872 Act Use Regulations Affecting Prospecting and Mining in National Forests," published September 1974.

Requirements

Here are the principal requirements of these regulations in brief form:

A person must file a "Notice of Intention to Operate" with the local Forest Service office in the event that any proposed prospecting or mining operation might cause "significant disturbance of surface resources." This notice is to include a plan of operation and steps the operator will take for rehabilitation of the area, and a bond to assure performance.

Without further explanation this may appear to be discouraging to the beginning prospector and recreational miner. Before forming a hasty opinion, please study the following interpretations given in question-and-answer form.

Interpretation

Interpretation of the meaning and application of this rather broad statement is important. Several question-and-answer statements found in the handbooks listed above are of special importance to any gold prospector. For example:

Question: What is meant by "significant disturbance" of surface resources?

Answer: In general, operations using mechanized, earth-moving equipment would be expected to cause significant disturbance. <u>Pick and shovel operations normally would not.</u> Almost without exception, road and trail construction and tree-clearing operations would cause significant surface disturbances.

Question: I'm a "rockhound" or mineral collector. How are my activities covered by requirements for operating plans or notices of intention to operate?

Answer: Your activities do not generally require either an operating plan or a notice of intention to operate. However, if you have any doubt about whether or not your activities will cause significant surface resource disturbance, you should file a notice of intention.

Question: Is an approved operating plan a form of permit?

Answer: No. Prospectors and miners already have express permission, by act of Congress, to go upon those National Forest lands which are open to operation of the 1872 mining laws for the purposes of "prospecting, locating and developing mineral resources."

Question: Do these regulations allow the Forest Service to tell me where and how I may prospect and mine?

Answer: No. They simply ask that you operate in a responsible way, with recognition of the need to protect non-mineral resources as much as possible.

Question: Will these regulations be fairly, consistently and reasonably administered by the Forest Service?

Answer: Yes. The Chief of the Forest Service committed himself to this in public speeches and congressional testimony. In addition, appeal procedures are provided for in the regulations if prospectors or miners believe they are being treated unfairly.

Question: Won't these regulations substantially reduce, if not completely halt, mineral exploration and development in the National Forests?

Answer: That is neither their design nor their purpose. They *will* substantially reduce unreasonable or unnecessary damage to surface resources.

Question: What is the single most important feature of these regu-
            lations for the purpose of minimizing the impact of pros-
            pecting and mining on the surface resources?

Answer: The prospector's and miner's operating plan. It provides a
            way for the mineral operator and the Forest Service to
            cooperate in developing workable procedures by which
            proposed operations can be done with the least impact or
            damage to both the surface resources and the operator.

Question: Exactly what is meant by an operating plan?

Answer: An operating plan, as required by these regulations, is a
            document by which a mineral operator identifies himself,
            describes the work he intends to do, where and when he
            intends to do it, the nature of this proposed disturbance
            of surface resources, and the steps he will take to protect
            those resources. An approved operating plan is basically
            an agreement between the Forest Service and the
            operators. The operator agrees to observe necessary and
            reasonable precautions, spelled out in this plan, to re-
            duce damage to surface resources during his activities
            and to rehabilitate disturbed areas when feasible. In turn,
            the Forest Service agrees that protection of surface re-
            sources will be adequate if operations are carried out in
            accordance with the approved plan.

Question: When is an operating plan necessary?

Answer: A plan of operations is required when any one whose
            proposed operations under the 1872 mining laws could
            cause "significant disturbance to the surface resources."
            An operator who is unsure whether his proposed opera-
            tions might disturb surface resources should file a "notice
            of intention to operate" with the Forest Service. It should
            describe briefly what he intends to do, where and when it
            is to be done, and how he intends to get himself and his
            equipment to the site. The Forest Service will analyze
            the proposal and within 15 days notify the operator as to
            whether or not an operating plan will be necessary.

Wilderness and Primitive Areas

As a gold prospector and potential miner you may be interested to know what your rights are in a wilderness or primitive area. The U.S. Forest Service is responsible for the administration of these National Forest Wilderness and Primitive areas.

Definition and Purpose

The purpose of a wilderness area is to "secure for the American people of present and future generations the benefits of an enduring resource wilderness."

The Wilderness Act of September 3, 1964 (Public Law 88-577) defines a wilderness as "An area where the earth and its community of life are untrammeled by man . . . an area of undeveloped Federal land retaining its primeval character and influence, without permanent improvements, protected and managed so as to preserve its natural conditions, with the imprint of man's work substantially unnoticeable. . . ." These areas ". . . shall be devoted to the public purpose of recreational, scenic, scientific, educational, conservational and historical use ."

Regulations

The wilderness areas were set aside by Congress in 1964. The act also provided that certain other areas in the national forests classified as "primitive" should be studied and reviewed to determine their suitability for preservation as wilderness. In other words, a primitive area is a candidate for wilderness. Their present status is a temporary one as they will eventually become wilderness areas to revert to National Forest.

Regulations relating to prospecting and mining are the same for both wilderness and primitive areas, but differ in some important respects to those effective in the National Forests. A rather complete source of information on these is found in Forest Service publication No. 275 entitled "Mining and Mining Claims in National Forest Wildernesses." This publication is available, free for the asking, at any Forest Service District Office. Following are several questions and answers taken from this publication which may suffice for your purpose:

Question: How do I know when I am in National Forest Wilderness?

Answer: Maps showing the Wildernesses in the various National Forests are available at the Supervisor's Office or Ranger Stations of the particular Forest or from the Regional Offices of the Forest Service. Also, the boundaries are posted along the trails leading into the Wilderness.

Question: May I enter a Wilderness and prospect for minerals?

Answer: Yes, until midnight December 31, 1983.

Question: What methods of prospecting may I use in my search for minerals in a Wilderness?

Answer: Most ordinary methods of prospecting are applicable, however, because of the effect of prospecting on the Wilderness, limitations have been placed on the size of excavations that can be made and controls are in force on the use of mechanized equipment. Prospecting permits are required where the limitations will be expected and/or mechanical equipment is expected to be used.

Question: When do I need a prospecting permit?

Answer: You must have a prospecting permit in order to do the following:

1. To use motorized equipment on the site or to transport it to the site by helicopter for prospecting use.

2. To dig one or more excavations which singly or collectively exceed 200 cubic feet within any area which can be bounded by a rectangle containing 20 surface acres.

3. To use overland mechanical means or helicopter to transport personnel to or from the mineralized site.

Question: At what stage of prospecting should I stake a mining claim?

Answer: The location of your claim should be made when you believe the evidence you have gathered—such as a mineral exposure, geologic indications, geophysical and/or geochemical information—indicates there is a reasonable chance that valuable mineral deposit is present at a particular site.

Question: How do I file a mining claim located in a Wilderness?

Answer: The staking and location procedure is the same as for other public land open to mineral entry, except that within 30 days after locating a claim you must file a written notice of the location of that claim and your Post Office address with the Forest Supervisor or District Ranger having administrative responsibility over the land on which the claim is located.

Question: What is meant by mechanical equipment?

Answer: Any equipment not powered by man or animal, except small contrivances as a battery-powered geiger counter, flashlight or radio.

Plan Ahead

From the above it appears the modern prospector, in most cases, will not need to file a notice of intention to operate, at least until he has a discovery developed to the point where he will be using mechanical equipment. By knowing the regulations and planning ahead, the smart prospector will probably find that he can carry on the work necessary for his discovery and subsequent assessment work without being in conflict with any existing regulations. When, and if, his development work requires singificant disturbance of the surface, he will find the Forest Service ready to cooperate with him in developing a plan that will permit development of the mine and protect the environment as well. The prospector should contact the Forest Service sufficiently ahead of his proposed activity to permit them to arrange time to discuss and work out the problem with him.

Wilderness areas are a heritage for all to enjoy.

The Federal Land Policy and Management Act of 1976 requires an affidavit be filed that the assessment work has been done.

Bureau of Land Management

Prior to 1976 the only record of unpatented mining claims, and assessment work done, was in the courthouse of the county in which the claims were located. Changes of ownership were often not recorded at all.

The Federal Land Policy and Management Act of 1976, passed by the 95th Congress, marked the beginning of a new era of public land management. This act outlined certain objectives concerning the recording of the location of all unpatented mining claims, and the filing of assessment work affidavits, or a notice of intention to hold such a claim.

Purpose

The Bureau of Land Management (BLM) of the U.S. Department of the Interior was charged with the responsibility of implementing these objectives on all public land administered by it. Accordingly, the BLM prepared regulations that became effective October 21, 1976, the stated purpose of which is to:

1. Establish a central record of the number and location of all unpatented mining claims, mill sites and tunnel sites located on Federal lands.
2. To maintain a record of the performance of assessment work done on each claim.
3. To provide a record of the transfer of interest of each claim.

Location

The new BLM regulations now require the owners of unpatented mining claims, mill sites or tunnel sites, located after October 21, 1976 on Federal lands, to file a copy of the original notice of location, filed in the courthouse, with the proper BLM office within 90 days after the date of location of the claim.

Location notices generally contain the date, name of locator(s), name of claim(s) whether the mining claim is a lode or placer, mineral(s) claimed, and the legal description by section, township and range, or by metes and bounds. If the state has no law requiring the recording of mining claims or sites, then the above enumerated information must be filed with the BLM within 90 days of the date of location. The above requirements also apply where the U.S. Government owns only the mineral rights.

Special Note (Claims located before October 21, 1976)

The owners of all unpatented mining claims, mill sites or tunnel sites located on or before October 21, 1976, had until October 22, 1979 to record their claims with the BLM. This has established, for the first time, a central record of all unpatented mining claims in the United States.

Assessment Work

In addition to the filing of the original location notice, the BLM also requires the filing of an exact copy of the affidavit of assessment work done on the claim, and recorded in the courthouse, or a notice of intent to hold the claim. This is to be filed in the District Office of the BLM before December 31 of the year following the year in which the affidavit was filed. Also to be submitted is the serial number assigned to each claim, and any change of address of the owners of the claim. If a notice of intention to hold is submitted, in lieu of a copy of the affidavit of assessment, it shall be in the form of a letter stating:

1. The serial numbers assigned to the claim by the BLM.
2. Any change of address of the owners of the claim.
3. A statement that the claim is held and claimed by the owners.
4. A statement that the owners intend to continue development of the claim.

5. The reason the affidavit of annual assessment work has not been filed, if it has not.

Transfer of Interest

To maintain a current record of the ownership of all unpatented mining claims, mill sites and tunnel sites, the BLM requires the transferee, or purchaser of the acquired interest in the claim, mill or tunnel sites, to file, within 60 days after the completion of the transaction, the following:

1. The serial number of the claim.
2. Name and address of transferee.

Any person who has acquired an interest in a mining claim, mill site or tunnel site by inheritance shall also file with the BLM the serial number of the claim and his name and address within 60 days after completion of the transfer.

Defend Discovery

In addition to the considerations above, the prospector who has made a discovery and filed his claim must be prepared to defend his discovery at any time. That is, he must be able to show that he has discovered a valuable mineral in a sufficient quantity to warrant its further development. This is a basic requirement of the mining law of 1872 and even though the Forest Service does not have the authority to enforce the requirement, they can, and do, check new claims for the validity of the discovery. If a claim shows no evidence of value they will then notify the Bureau of Land Management, which does have the authority to contest and cancel the claim.

Question: How do I file a mining claim located in a wilderness?

Answer: The staking and location procedure is the same as for other
            public land open to mineral entry, except that within 30 days
            after locating a claim you must file a written notice of the loca-
            tion of that claim and your post office address with the forest
            supervisor or district ranger having administrative responsibil-
            ity over the land on which the claim is located.

Knowledge, born of experience and only shared with other old timers, produced a reservoir of secrets not found in any textbook. The campfires where tall tales and true adventures were told were the prospector's trade schools and seminars. Wherever gold prospectors met, the talk was about how and where to find gold. Although the old gold mining days are history, the secrets of the old prospectors have lived on. This book will tell you about these colorful men and the mining skills they used in the days when gold mining was a way of life and a crucial economic venture.

Bureau of Land Management (BLM) District Offices in The U.S.

ALASKA—SOUTHERN
Anchorage Land Office
555 Cordova Street
Anchorage, Alaska 99501

IDAHO
Idaho Land Office
Federal Building
Boise, Idaho 83701

ALASKA—NORTHERN
Fairbanks District and Land Office
516 Second Avenue
Fairbanks, Alaska 99701

MONTANA
Montana, North Dakota and South Dakot
Granite Building
Billings, Montana 59101

CALIFORNIA—NORTHERN
Sacramento Land Office
Federal Building
Room 4017
Sacramento, California 95814

NEVADA
Nevada Land Office
Federal Building 300 Booth Street
Reno, Nevada 89502

CALIFORNIA—SOUTHERN
Riverside District and Land Office
1414 University Avenue
Riverside, California 92507

NEW MEXICO
New Mexico, Oklahoma & Texas
New Mexico Land Office
New Federal and U.S. Post Office
South Federal Place
Santa Fe, New Mexico 87501

COLORADO
Colorado Land Office
Federal Building 1961 Stout Street
Denver, Colorado 80202

OREGON
Oregon and Washington
Oregon Land Office
729 Northeast Oregon Street
Portland, Oregon 97232

EASTERN
Arkansas, Iowa, Louisiana, Missouri
Minnesota (minerals only) and all
states east of the Mississippi River.
Eastern States Land Office
7981 Eastern Avenue
Silver Spring, Maryland 20910

UTAH
Utah Land Office
Federal Building
Salt Lake City, Utah 84111

WYOMING
Wyoming, Kansas and Nebraska
Wyoming Land Office
U.S. Post Office and Courthouse
Cheyenne, Wyoming 82001

United States Forest Service Regional Offices

ALASKA
P.O. Box 1628
Juneau, Alaska 99801

PACIFIC NORTHWEST
P.O. Box 3623
Portland, Oregon 97208

CALIFORNIA
630 Sansome Street
San Francisco, California 94111

ROCKY MOUNTAIN
Federal Center Building 85
Denver, Colorado 80225

INTERMOUNTAIN
324 Twenty Fifth Street
Ogden, Utah 84401

SOUTH
Suite 800
1720 Peachtree Road N. W.
Atlanta, Georgia 30309

NORTH
Federal Building
Missoula, Montana 59801

SOUTHWESTERN
517 Gold Avenue S. W.
Albuquerque, New Mexico 87101

Conclusion

Now that we have told you the secrets of the old '49ers they are no longer secrets in one sense of the word. *Webster* defines a secret as information "hidden from others" or "revealed to none or to a few" so perhaps we can still say they are secrets since we have revealed them to only a few; to you and a few others. This is still only a small number, considering how many prospectors and miners are going to be around looking for gold in the next few years. In any event, we have enjoyed passing on some of the old timers' secrets to you and hope they may be helpful.

Bill Little, my partner, is the one who really contributed most of the secrets. He spent over 25 years up in the High Sierra gold country in northern California after he got out of the Air Corps in 1945. He knew most of the old timers who remained up in that country, and he learned a lot from his own experience. You can't beat experience. When you get to checking out those outcrops or tiltin' a pan on some likely creek you'll find that you will soon be your own teacher, doing what comes next, and naturally. That's experience!

After Bill told me some of the old timers' theories I decided to check some of them out in the books. I mean the geology books. I read several of them and found that those old timers were pretty well up on their geology—even though they probably couldn't even spell the word.

One day, Bill and I got to talking, and I said "Bill, why don't we write a little six or ten page pamphlet on how to find gold? We'll mimeograph it and maybe sell a few." He said, "It's all right with me,

135

Many of the old and forgotten mines will again be opened and made to produce their gold.

if you do the writing." Well, that started the ball rolling: I found that six or ten pages weren't enough to hold all the old timers' secrets that Bill came up with, so we had to make it larger. One thing led to another, until the pamphlet turned into a booklet, then into a book and that is what you have in your hand.

Bill told me to be sure to wish you a lot of luck in your prospecting—he said you would need it. I'll join with him in the wish, so we both say, "GOOD PROSPECTING, PARDNER," and if you strike it rich from anything you learned out of this book we would sure be glad to hear about it. Just drop us a line in care of our publishers and they will pass it on.

Verne H. Ballantyne

Glossary

Definition of Mining Terms

(Courtesy Montana Bureau of Mines and Geology)

ADIT   A horizontal gallery or opening driven from the surface of the ground to the ore body. The term "tunnel" is frequently used in place of adit, but technically a tunnel is open to the surface on both ends.

ALLUVIAL   Generally pertains to loose gravel, soil, or mud which has been deposited by water.

ANALYSIS   A separation of compound substances by chemical means.

ASSAY   The determination of the valuable minerals in a sample. A wet assay is determined by the use of chemicals. A fire assay is determined by both chemicals and fire. Gold and silver are usually assayed by fire.

BEDROCK   Any solid rock underlying gold-bearing gravels.

BLACK SAND   Grains of heavy, dark minerals such as magnetite, limenite, chromite, etc., found in streams which commonly collect in sluice boxes and which may carry gold and platinum.

CHUTE   An opening in the ground where ore is allowed to pass

137

from one level to another. It is the structure built to load cars from a stope.

CLAIM   A land area claimed by a prospector and marked out by stakes.

COLOR   A term referring to small grains or flakes of gold.

CONTOUR   Lines connecting points of equal elevation on a contour map.

DIP   The maximum angle of inclination downward that a vein or bed makes with a horizontal plane.

DYNAMITE   Dynamite is an explosive mixture of glycerin, sodium or ammonium nitrate, and a filler of combustible pulp such as a wood meal.

ELECTRIC CAP   A small metallic cap containing fulminating powder which is detonated by an electric current.

EXPOSURE   Any part of a vein or rock outcrop that can easily be seen.

FAULT   A fracture in the earth, with displacement of one side of the fracture with respect to the other.

FISSURE   An opening or crack in the rock. A fissure vein is a fissure filled with mineral matter.

FLOAT   The loose and scattered pieces of ore which have been broken off from an outcrop.

FOOTWALL   The bottom or lower enclosing wall of a vein.

FUSE   A tube or cord filled or impregnated with combustible matter for igniting an explosive charge after a predetermined interval, as in blasting.

HANGING WALL   The top or upper enclosing wall of a vein.

HEAD FRAME   A structure erected over a shaft to support the sheave wheel for hoisting purposes.

HEADING   Any part of a mine where work is under way. Usually confined to development workings only.

HIGH GRADING   Stealing of high grade ore or nuggets from the workings of a hard rock or placer mine by employees or others.

IGNEOUS ROCK   Rock formed from molten lava.

LATERAL   A horizontal mine working. A drift in the footwall of a vein is often called a lateral.

LEASE   A contract by which one conveys real estate for life, for a term of years, or at will, usually for a specified rent or royalty.

LESSEE   A person who obtains a lease on mining land.

LESSOR   The grantor of a lease.

LEVEL   All the connected horizontal mine openings at a certain elevation.

LOCATING   The marking of the boundaries and staking of a mining claim.

LODE   Refers to a tabular deposit between definite walls.

MILLING ORE   Ore that must be concentrated at or near the mine before it is shipped.

MUCK   A common term for any broken ore or underground waste.

MUCKER   A shoveler, or one who handles muck.

NITRO   Short for nitroglycerin, which is any nitrate of glycerol, a colorless, heavy, oily, explosive liquid used in making dynamite.

NUGGET   A piece of gold of any shape or size larger than a flake, usually rounded by stream and water action.

OPTION   This is the right to purchase at a stated price.

ORE   A mineral aggregate of sufficient value to be mined at a profit.

ORE BODY   The part of a vein that carries ore. Generally, all parts of a vein are not ore. Ore shoot has the same meaning.

OUTCROP   The edge or surface of a mineral deposit or sedimentary bed which appears on the surface.

OVERBURDEN   The valueless material overlaying the pay zone in a placer deposit or the waste or valueless material of a solid outcrop.

OXIDE   A compound of a metal and oxygen.

PATENT   A written title to land granted by the government after meeting certain obligations. A mining claim can be patented after $500 worth of work has been done and other requirements met.

PLACER   Alluvial deposit of valuable mineral-bearing gravel.

PLANE   An even surface. A horizontal plane is a flat, even, level surface.

POWDER   A miner's term for dynamite or other explosive.

RAISE   An excavation of restricted cross-section, driven upwards either vertically or at an angle from a level in the mine.

RAKE   The trend of the ore body within the vein.

RIFFLE   Grooves, channels, slats, or wire screens in a sluice box or rocker to catch gold or other valuable minerals.

SLIP   A small fault.

SLUICE BOX   A trough paved with riffles through which gravel and wash from placer mining operations pass so that gold and other valuable minerals can be caught and saved.

SPECIFIC GRAVITY   The ratio of the weight of any substance to the weight of an equal volume of water.

STOPE   Any excavation underground used to remove the ore.

STRIKE   The bearing of a horizontal line in the plane of a vein, bed, or fault in respect to the cardinal points of the compass.

STRIPPING   Removal of the overburden from a placer deposit or the barren outcrop from an ore deposit.

STULL   A timber used to support loose rocks or slabs. It may also be used to support a platform in a working area.

TREND   The general direction or bearing of a vein, fault, or rock outcrop.

VALUE   Refers to the mineral substance searched for. In the case of gold the term is synonymous with color.

VEIN   A well-defined, tabular, mineralized zone which may or may not have valuable ore bodies.

WALL   The waste or country rock on either side of a vein.

WASTE   Barren rock or mineralized material which does not have enough value to be classified as an ore.

WORKING FACE   Any portion of the mine where work is under way, such as the face of a drift or the face of a raise.

Bibliography

Books

Bacon and Beans From a Gold Pan. Jesse L. Coffee; George Hoeper, Doubleday & Co., Garden City, New York

An account of the personal experiences of the author and his wife during the depression years of the 1930's in the Mother Lode placer areas in California, where they made their living working the gravel bars in the streams of the High Sierras.

Blaster's Handbook. E. I. du Pont de Nemours & Co., Wilmington, Delaware

A practical and complete manual giving instructions for setting up a charge, both electric and fuse; quantities of dynamite for different hardnesses of rock; safety precautions; and many details the powder man and amateur miner or prospector should know.

Diving Digging For Gold. Mary Hill. Naturegraph Publishers Inc., Herldsburg, California

Brief, 47-page discussion on recreational mining from a historical perspective.

Gold Mining for Recreation. Harvey Neese, Chronicle Books, San Francisco, California

This 78-page book on recreational prospecting and mining for gold was published in 1981.

Gold! The Way To Roadside Riches. Tom Bishop. Johnson Publishing Co., Boulder, Colorado

This is a 59 page study on recreational, hare fock prospecting. Prospecting with the aid of electronic, treasure hunting equipment is discussed.

Gold Placers and Placering in Arizona. Arizona Bureau of Mines, University of Arizona, Tucson, Arizona

This 123 page descriptive booklet gives information on known placer locations in Arizona along with methods and equipment used in placer mining.

Handbook For Prospectors. Richard M. Pearl. McGraw Hill Book Co., New York

This is a modernized and abridged version of the Von Bernewitz book titled *Handbook for Prospectors and Operators For Small Mines*.

Handbook For Prospectors and Operators For Small Mines. W. M. Von Bernewitz, revised by Harry C. Chellson

This large, 531-page, hardcover handbook is out of print, but may be available in your library. It contains much valuable mining information, is somewhat technical, and some of the material and pictures are out of date. It was originally copyrighted in 1926 but has been revised and the copyright was renewed in 1963.

SPECIAL NOTE: An up-to-date, fifth edition of the *Handbook for Prospectors*, based on the Von Bernewitz book and edited by Richard M. Pearl, is now available from McGraw-Hill Book Co., 1221 Avenue of the Americas, New York, N.Y. 10020. It is 596 pages, 5½" × 8", and sells for $14.50.

How and Where To Prospect For Gold. Verne H. Ballantyne. Tab Books Inc., Blue Ridge Summit, Pennsylvania

This book has 240 pages and over 100 illustrations. It is written in layman's language for the beginning or recreational gold prospector. It is a practical treatise of the methods and equipment used in the search for placer gold primarily but an overview of hard rock prospecting is given. Processing the concentrates, marketing the gold, government regulations, staking a claim and the business end of prospecting are all discussed in an interesting and informative manner. One of the most complete of the gold prospecting books.

How to Build and Use Your Own Mini-Rocker. Verne H. Ballantyne. Ballantyne Publications, P.O. Box 477, Bozeman, Montana

Complete instructions, bill of materials, working drawings and detailed information on how to build and use a mini-rocker are given in easily understood layman's language. Gold prospecting in the simplest terms. Soft cover, 22 pages, zeroxed special studies edition.

How To Make Prospecting For Gold A Business. Verne H. Ballantyne. Ballantyne Publications, P.O. Box 477, Bozeman, Montana

The business aspects of prospecting for gold are developed in this booklet. Gridiron sampling to obtain an estimate of values, how to determine the value of a gold property, necessity for written agreements, problem of theft, leasing, selling or operating problems are all discussed. Soft cover, 19 pages, zeroxed, special study edition.

How to Process Your Black Sand Concentrates. Verne H. Ballantyne. Ballantyne Publications, P.O. Box 477, Bozeman, Montana

A step-by-step method of processing the black sand concentrates is given in this up-to-date booklet. Both a sample size quantity and commercial amounts are discussed. Both methods and equipment are stressed. Soft cover, 22 pages, zeroxed, special studies edition.

How To Prospect For Hard Rock Gold. Verne H. Ballantyne. Ballantyne Publications, P.O. Box 477, Bozeman, Montana

Both modern and old-time methods of prospecting for hard rock gold are revealed. How to find the Mother Lode is described in detail. Explanations of theories of gold formation, relation to mountain building and most likely areas to prospect. Common terms defined and explained. A practical handbook for the beginning hard rock prospector.

Soft cover, 20 pages, xeroxed, special study edition.

How To Prospect For Pocket Gold. Verne H. Ballantyne. Ballantyne Publications, P.O. Box 477, Bozeman, Montana.

This little known kind of gold deposit even by the old time prospectors is discussed in detail and the secret method of tracing them down is told. Information obtained direct from an old time pocket hunter.

Looking For Gold—The Modern Prospector's Handbook. Bradford Angier, Stackpole Books, Harrisburg, Pennsylvania

Gives a brief description of the placer potential of the western states and Canada. Basic prospecting and placer mining information is given together with suggestions on leasing, grubstaking and selling mining property.

Placer Examination—Principles and Practice. John H. Wells, Superintendent of Documents, Washington, D.C.

This hardcover book has excellent information on the use and construction of rockers, sluice boxes and mechanical gold washing machines with numerous illustrations and photographs.

Practical Guide For Prospectors and Small Mine Operators in Montana. Koehler S. Stout. Montana Bureau of Mines and Geology, Butte, Montana.

A 103-page softcover booklet with six pages of drawings and sketches, which contains excellent information on prospecting, estimating, developing, mining methods, placer mining, Montana Mining Laws, Revised Codes of 1947; a table of chemical elements and a list of references; and lists of assayers, consulting engineering firms, machinery suppliers, and miscellaneous mining and equipment suppliers. It is in print and a bargain at $1.00 per copy.

Principle Gold Producing Districts of the United States. A. H. and Bergenhahl, M. H. Koschmann and M. H. Bergenhahl. Superintendent of Documents, Washington, D.C.

This publication gives a thorough descriptive listing of all the gold mining districts in the United States with production data and geologic information.

Prospecting and Operating Small Gold Placers. William F. Boericke, John Wiley & Sons, New York

This long time favorite covers all phases of placer mining.

Shallow Diggins—Tales From Montana's Ghost Towns. Jean Davis, Caxton Printers, Ltd., Caldwell, Idaho

A collection of historical narratives of people and events in the various mining camps and settlements in early Montana. Out of print but available in libraries.

Simple Methods of Mining Gold. Terry R. Faulk. The Filter Press, Palmer Lake, Colorado

Oriented to recreational placer mining. Describes specific areas where gold is found in each of the gold-producing states.

Small Fortunes in Penny Gold Stocks. Norman Lamb. SRT Corporation, P.O. Box 148, Vallejo, California

The basic theme of this interesting book is how to make money by buying penny gold stocks. The author appears to be knowledgeable on the practical side of mining and geology as well as in the skill and timing of the market. The book gives much valuable information on the economic and business end of gold and silver mining.

Survival With Style. Bradford Angier, Stackpole Books, Harrisburg, Pennsylvania.

This book is a complete source of information of how to survive in the forest, desert, wilderness or other mountain country. Developed for National Wildlife Federation, Washington, D.C.

Treasure's Hunter's Guide (How and Where to find it). Robert I. Nesmith and John A. Potter. Arco Publishing Inc., New York.

Discusses all types of treasure hunting. The section on skin and scuba diving is full of information and encouragement for the underwater crevice miner who wishes to use floating dredge equipment.

The World of Gold. Timothy Green. Walker Publishing Company, 1973, New York, N.Y.

A general treatise on gold: the history of its use, mining, and political and monetary considerations.

OTHER PUBLICATIONS

American Gold News, P.O. Box 457, Ione, California 95640.

A monthly publication containing many articles and news of interest to the gold prospector. Deals entirely with gold-related information and political developments.

California Geology, P.O. Box 2980, Sacramento, California 95812.

A monthly publication issued by the California Division of Mines and Geology. It contains excellent and timely articles on many phases of mining of interest to the prospector and small miner.

California Mining Journal, 2539 Mission Street, P.O. Drawer 628, Santa Cruz, California 95061.

This is the main trade journal of the small mining industry of the western states. Originally principally oriented to California has become of regional and national interest. Articles are timely and comprehensive dealing with any and all aspects of mining with emphasis on gold prospecting and mining.

Denver Equipment Co., Division of Joy Mfg. Co., 1499 Seventeenth St., Denver, Colorado 80217.

Publishes various bulletins and articles on gold production in addition to sales information and equipment available for small mining operations.

Engineering and Mining Journal, McGraw-Hill Building, 1221 Avenue of the Americas, New York, N.Y. 10020.

This is the leading trade journal of the mining industry in the United States. Oriented to commercial and large scale mining. Arti-

cles are of national interest to miners and those associated with mining. Published monthly.

Gold Districts of California, Bulletin No. 193, Published by the
California Division of Mines and Geology, Ferry Building, San Francisco, California 94111.
   Detailed description of principal gold-bearing districts in California.

Gold Placers of California, Bulletin No. 92. Charles Scott Haley.
Published by Division of Mines and Geology, Ferry Building, San
Francisco, California. 94111.
   Gives brief descriptions of the location and extent of the principle
gold placers in California, including maps. The author, Charles Scott
Haley, was a prominent, private consulting mining engineer, en-
gaged by the California State Mining Bureau to conduct the necessary
field work and prepare this valuable publication. Written in 1923 after
two years of intensive field research. It is now out of print and available only in libraries. Contains much historical information on the
problems arising from hydraulic mining.

Gold Prospector's News, Gold Prospector's Association of America,
P.O. Box 507, Bonsall, California 92003 (mailing address) and 2605
Buena Rosa, Fallbrook, California 92028 (office address).
   This is the official publication of the Gold Prospector's Association
of America. It is published every other month and distributed to
members as part of the benefits of membership in the GPAA. Contains timely information and news relating to recreational prospecting
and mining for gold in all parts of the United States and Canada. The
GPAA is the largest prospecting organization in America with a membership of over 60,000.

How To Mine and Prospect for Placer Gold, Bureau of Mines Information Circular No. 8517, By J.M. West. Available from the Superintendent of Documents, Washington, D.C. 20402.
   This publicaiton gives a brief history of placer mining in the United
States. Discusses where to prospect by area and state. Extensive
bibliography.

Laws and Regulations Covering Mineral Rights in Arizona, Department of Mineral Resources, State of Arizona, Phoenix, Arizona.

This is a booklet containing the laws and regulations relating to mining in Arizona. Essential for anyone expecting to stake a claim in Arizona or interested in mining in that state.

Legal Guide for California Prospectors and Miner, Special Publication 40, California Division of Mines and Geology, Ferry Bldg., San Francisco, California 94111.

An invaluable source of information on all legal phases of mining in California. It is the most popular item published by the Division of Mines and Geology.

Locating Gold, United Prospectors, Inc., 5665 Park Crest Drive, San Jose, California 95118.

This is a magazine type of publication published every two months. The United Prospector is a recreational type of organization and the publication is one of the benefits of membership.

Patenting a Mining Claim on Federal Lands, Bureau of Land Management, U.S. Department of the Interior, Washington, D.C.

This small pamphlet provides basic information on how to file an application for a patent to a mining claim. The requirements and the administrative procedure involved. Free upon request.

Pay Dirt, Pay Dirt, P.O. Drawer 48, Bisbee, Arizona 85600.

This is the official publication of the Arizona Small Mine Operator's Association. Contains many articles on various phases of mining in Arizona, including political news related to mining. Published monthly.

Placer Mining for Gold in California, Bulletin No. 135, Charles Volney Averill, Published by the California Division of Mines and Geology, Ferry Building, San Francisco, California 94111.

This valuable bulletin on placer mining in California is now out of print but can be obtained from libraries.

Staking a Mining Claim on Federal Lands, Bureau of Land Management, U.S. Department of the Interior, Washington, D.C.

Provides in question-and-answer style, the fundamental facts on where and how to stake a mining claim. Discusses size and shape of lode and placer claims and other important information. May be obtained from any Bureau of Land Management Office or from the Superintendent of Documents, U.S. Government Printing Office, Washington, D.C. 20402.

Western Prospector and Miner, Western Prospector and Miner, P.O. Box 146, Tombstone, Arizona 85638.

This is a newspaper-style publication containing many articles and news items of interest to the gold prospector and miner. Published monthly.

Index